HOW TO
BEAT THE
SYSTEM

HOW TO BEAT THE SYSTEM

The Student's Guide to Good Grades

KATHY CRAFTS
& BRENDA HAUTHER

GROVE PRESS INC., / NEW YORK

First Black Cat Edition 1982
First Printing 1982
ISBN: 0-394-17740-1
Library of Congress Catalog Card Number: 81-47643

Manufactured in the United States of America

GROVE PRESS, INC., 196 West Houston Street, New York, N.Y. 10014

So smile while you're makin' it,
Laugh while you're takin' it,
Even though you're fakin' it,
Nobody's gonna know. . . .
————Alan Price*

* From *O Lucky Man!* copyright © 1973 by
Keith Prowse Music/Jarrow Music

We would like to express our thanks to:

KATE STIMPSON . . .
Professor Summa Cum Laude
Cassie Stromer
Karen Hansen
DBLT Hauther
Barbara Nichols
Nis Lyons
Bernie Naskret
Samuel Ready
Diana Driscoll
E.M.
Kathy Doheny
Sylvia McDonough
Blake Murphy
Suzanne Ray
Richard Snee
Suzyn Jurist
Merri Spear
Laura McQuinn
Ronda Wist
Patricia Golin
David Niven
Errol Flynn

All the other people at Columbia University

CONTENTS

PREFACE

The academic scene has changed radically in the six years since this book was written. Six years ago, business schools were considered "walk-ins." Anyone could get an MBA. Just about anyone with decent grades and cash could get an Ivy League MBA. A student scoring 590 on the LSAT could go to law school—not one of the better ones, to be sure, but a fairly good one. Some American students, unable to get into medical school here, applied to foreign universities but this was not considered a viable alternative. Just a few short years ago, when we were undergraduates, it was possible to wipe out a poor record with a good board score and then trip off merrily to graduate school. Obviously the situation is very different today.

More people than ever before are seeking admission to graduate and professional schools. At the same time, the Freedom of Information Act and the Truth in Testing laws have forced the testing services out of the closet. The value of the grad boards is being questioned. Dynamite test scores no longer ensure admission. Committees are now placing more emphasis on a student's overall record—where it should have been all along. But with this new emphasis it becomes more apparent that anyone who wants to get ahead must be aware of what is going on from the very first moment he or she walks into undergraduate school.

Ignorance is hardly bliss in this situation. Our intention in writing this book was to clue in high school seniors and college freshman. Sure, it's impossible to know if anyone really intends to go on to med or law school when he is only seventeen. But it would be a shame if that person should, through simple ignorance of the system, be rejected by every medical school in the world

i

five, eight, or ten years later. This happens. A freshman's D in Organic Chemistry follows him—they mark it in dayglo red on the transcripts. No one tells the kid that everyone audits this course before taking it for a grade. We intended this book to provide such information, or "instant academic street smarts."

Well, it was inevitable that many would be angered by what they read here. Everyone knew what we said was true, but quite a few people felt that none of it should be written down, much less published. We were accused of being anti-intellectual. Apparently it is far better for a student to land a few Ds than to take selected courses pass/fail and ensure the possibility of continuing his education beyond the undergraduate level. Our favorite review read, "A book for people who intend to read nothing longer than a stop sign after they leave college." Obviously this reviewer hadn't been on a college campus or applied to any sort of school in the last ten years. This book was and still is intended for those who do plan to read and might like to do so in a graduate or professional school someday. At the very least it is intended for the student who appreciates the value of keeping his options open. But ignorance of the system can close off most of these options in the opening moments of a college career.

Most students today know ignorance kills. They just don't know what to do about it. Most freshmen don't even know whom to ask. They pick up a tip here or there, but it isn't all laid out in a *Barron's Guide,* yet. That is the purpose of this book—to lay it all out and show where the dangers lie.

We went on to grad school; one lawyer, one MBA. Not everyone will go, nor should everyone. But we still feel that everyone, even the lowliest freshman, deserves to know all the rules of the game—before he or she seals his fate.

—K.C.
B.H.

INTRODUCTION

The art of self-defense, as any judo expert will tell you, is using your opponent's strength to work for you. So it is with the educational system at the university level. The strength of its inertia will break you if you try to buck it. The secret is to use the system to your advantage. You are not going to change it (as evidenced by the failure of the student actions in the '60s), so you might as well learn to manipulate it and make it work for you.

The problem this philosophy presents is that the educational system absorbs many people every year who have absolutely no knowledge of it. By the time these people wake up, it is too late. In this book, we attempt to solve this problem. As graduates and triumphant survivors, we can tell you what we wish we had known, and what we learned through painful trial and error. We can tell you what we know works and what we know doesn't. Like experienced hunters (and let's face it, undergraduate school today is a competitive jungle), we can warn you about traps and pits, steer you away from tempting but deadly lures, and give you a short course on how to become an academic guerrilla.

The only thing a freshman lacks is experience: the type of experience and perspective a senior has when attempting to get into a graduate school. After four years he knows *how* to write a paper, he knows *how* to wheedle a professor for a better grade, he knows *how* to take an exam, he knows *how* to look at a reading list and discover which books are important, but he is still carrying the scars of those first two or three terms when he did not know. And ironically enough, the courses that he screwed himself to the wall with were most likely not even in his major.

To prove just how essential some basic training is, we offer this too sadly true tale of a naive freshman:

Eager to get a start in her premed major, our heroine —a first-term freshman—registered in organic chemistry, which is the well-known make-or-break course for medical school. Unfortunately, this freshman was unaware that many of the upperclassmen had sat in on the class for several terms before taking it for a grade. She was unaware of the system of course dropping, pass/fails, extensions and the entire system of college grading. Had she been aware of these she might have been able to get out before it was too late. She did not know however and at the end of the term, this freshman, who might have been a finer doctor than anyone in her class, had gotten a "D" and with that "D" had virtually destroyed what might have been a fulfilling career in medicine.

Rather naive, you might say, but this has happened to more than one student in recent years. You might even have been one yourself if you had not been so zealous as a freshman, OR IF ONLY SOMEONE HAD TOLD YOU. If only someone had told you that people in college do not operate by the rules of high school, you might still be in the running for the graduate school of your choice. So you may know just how invaluable a knowledge of the system is. Unfortunately, you know it now, when it is too late.

The first thing a new student must realize is that college is not like high school. It is like nothing he has ever experienced before: it is the ultimate bureaucracy, and like any bureaucracy it can be manipulated. Most freshmen are woefully unaware that when they first walk through the ivy gates, they need not be helpless. Talk about a babe in arms! He is leaving the playground and with minimal knowledge and experience going into the academic war. But with a little basic training he can come out ribbons flying, all shining glory. You see, unlike high school, there is no reason for getting terrible grades in college. In high school, many subjects are shoved down your throat regardless of your aptitude or interest. (All the shoving motivated by the revered theory that you never know what hidden delights and aptitudes you might find in the young high-school pupil.) But in college you almost never have to take courses that do not interest you and the few that you may have to take, you never have to take until you are damned good and ready. In those courses you absolutely cannot stand, you do not have to get miserable grades. Unlike high school, the highest grade does not always go to the one who knows the most—there are too many people for the professors to find out just who that person is. There is no reason why, with a minimal amount of effort and study, you should not get at least, at the very least, a "B" average.

Coming into this wide world from the narrow confines of high school or prep school can be momentarily confusing, but you can adapt. You must gain perspective. It may not be easy to adjust, but it is essential. The frame of reference becomes so much larger that you have to expand your horizons. Comparing high school to college is like trying to compare a white Kleenex to an intricately woven seventy-foot-high tapestry. You simply cannot. There is such scope and depth in the tapestry that the Kleenex gets blown away with the wind. In tune with this perspective, you must understand that perhaps the major difference between high school and college is the approach to teaching. Teachers in high school place the emphasis on the teaching, and not on the subject. In college, professors tend to think of themselves as art historians, scientists, writers, etc.—in short, professionals who are giving the benefit of their expertise. They are genuinely and usually passionately involved with their subject. Most of them regard the registrar, the deans, and grades as rather tiresome and troublesome: evils, to be sure, but necessary evils. Your grades and how you personally are doing are not first and foremost in their thoughts when they walk into that classroom. This is not to say that professors take no personal interest in their students; of course, many do. But you can bet your Bic Banana that they are more interested in you the person than in you the pupil. They feel no compulsion to make you taste the fruit of knowledge. For, unlike your high school teachers who may have seemed obsessed by the need to force-feed you, they are merely going to show you the tree.

This unexpected release of pressure has had disastrous effects on freshmen. Some spring wildly away and cause irreparable damage by their initial carelessness. Others, determined to handle their freedom maturely, are overzealous, get in over their heads, and wreak havoc that way. Still others, dazed by the whole thing, just screw up unintentionally. So we merely warn you that it is what you do *not* know that can kill you.

All this talk of learning the system and learning how to manipulate it may offend many people, especially the more serious-minded students. But even the most forthright can get shafted if they do not know what they are up against. Nevertheless, some people may be appalled by the seemingly anti-intellectual attitude of this book. That is not what we intend. It is meant to be a survival manual; the competition for grades and the small number of places in choice graduate and professional schools has managed to kill almost all intellectual curiosity among undergraduates these days. Students are afraid of taking anything they are interested in outside their major fields and slightly beyond their aptitudes because they fear ruining their averages. Competition among majors in upperclass levels is at the cutthroat level. It is for the chosen few blessed by the gods that the college course is uninterrupted smooth sailing. And while this book is dedicated in awe to their luck, most of us have to make our own luck and this book will help you. Remember, ignorance of the system is the kiss of death.

In essence, then, the purpose of this manual is to help you get grades in your courses that will boost your average, rather than drag it down. It should help you past the rough spots, and give you a little more of the perspective that only a rejected senior has. You need not make our mistakes. IF ONLY SOMEONE HAD TOLD YOU need never be your lament. If only someone had told us many of these tricks of the academic war five years ago, we might have made it into law school rather than writing this book.

GETTING IN, OR YOU CAN GO ANYWHERE

There are a couple of thousand colleges in this country, yet when you are working on your applications it seems as if there will be plenty of room for everyone except you. Guidance counselors aren't much help giving much beyond the traditional advice (i.e., apply to a dream school, a few you are almost sure of getting into, and a safety school), and then providing directions to the nearest state university. No one ever admits that it isn't a question of a school accepting you, but rather a matter of deciding where you want to go. Yes, friends, it's true! If you are willing to make an effort and use a few tricks of the trade, you can go to any school you want to. It's all a matter of strategy.

As you sit there, with the counselor's words thundering in your head and surrounded by applications, we know you don't believe a word of this. (Well, you've got the first lesson down. Don't believe everything you read.) Glancing through any newspaper today, you get the impression that the freshman class at an "elite" school is made up of superbeings. Not true! All eye-witness accounts indicate these freshmen are average seventeen–eighteen-year-old humanoids, just like you. Remember, the people interviewed by the press were handpicked by the administration to impress the paper, and in turn the world. The rest of the class (generally some five hundred to one thousand others) have not grown up in Outer Mongolia, did go to normal high school, do not read ancient Greek or Egyptian, and do prefer rock to opera. There is no reason why you shouldn't be one of them.

Okay, let's get down to brass tacks. You have assessed your financial position, grades, and test scores, and have narrowed it down to three or four schools. The guidance counselor has been some help, but not much. In most cases the counselor can be counted on to come up with the safety school, and to throw a little cold water on your dreams of attending one of the better universities. You mustn't take this seriously. Counselors always think people aim too high when applying to colleges. So take everything the counselor says with a grain of salt. Chances are good that he vividly recalls the disaster of his career—the student who didn't get in anywhere.

Now let's look at the schools you have chosen. Are you sure you aren't aiming too low? Most people do, and a good deal of the selection process goes on before the first application is mailed. Having SATs slightly below averages published in the *Barron's Guide* doesn't mean there isn't a chance of getting into a certain college. To begin with, these are *averages*. Also, you don't know what these schools are looking for at any particular time. One year, an "elite" school in New England faced a crisis: it seems most of the marching band would be graduating in the spring. The following fall, many of the freshmen were musicians with marching experience. Other schools, seeking to maintain a "national reputation," are extremely selective in regard to local students but actively recruit out-of-staters. In other words, if you apply to a "national" school, you will be competing not against 8000 people for 1000 seats, but against the four or five Californians applying for the two or three California freshman slots. Keep this in mind. Due to the emphasis on "geographical balance," competition decreases in direct proportion to the increase in distance. If you are willing to travel, you can probably go to a better school than the best you could get into in your own state. Don't let the counselors dissuade you. Just because no one from Salt Lake High has ever gone to college in Vermont is no reason for you to forget about applying to Middlebury. (If no one ever left home, there wouldn't be an America—lay that one on them.)

If you are dreaming of an incredibly selective univeristy and feel that, no matter what, you just can't get in—don't give up. A university is a collection of colleges. While you can't get into the College of Liberal Arts, you may be able to get into the School of Agriculture. Now you may be wondering what good is this— you don't want to be a farmer. Well, once you are in any one of the many colleges, you can transfer within the university. You don't have to lose time or credits even if you are registered in the School of Agriculture. You see, one advantage of attending a university is the fact that you can take courses in any of the colleges thus becoming a well-rounded person. The first term you can register for the same courses a freshman in the liberal arts college will be taking. You do the same thing every term until your "intra-college" transfer is complete. (Such a transfer is very common and very easy. More often than not, people transfer from liberal arts to more specific areas. So don't worry, you don't have to be a farmer.) The point is, you simply "find yourself" about two years down the road and "realize" you should be in the liberal arts school or whatever division your heart desires. The same sort of thing can be done within a university with a branch campus. Two terms down the road, you just apply for a transfer to the well-known and incredibly wonderful main campus.

In regard to the actual application, there are a few things to keep in mind. First and foremost, it should be well written. If you essay is sprinkled with typos and grammatical errors, forget it. You might as well send the school a donation rather than an application fee check. Grit your teeth and take the application to your English teacher. (A guidance counselor may have some good ideas about what you should say, but he is no expert on how you should say it.) This is no joke. The fact that you were one of the measly hundred out of three thousand applicants who could write a coherent essay can put you at the top of the heap, even if your test scores stink.

Aside from presenting a well-written essay, there are other ways to separate yourself from the thundering herd. Most students require financial aid. Today most schools have more than 50-60% of their students on some sort of aid program. You may need aid too (everyone does, no one can afford college anymore), but when applying to your dream school, you shouldn't say you do unless the school has a policy of withholding that information from the Admissions Committee. Given these hard times, schools are looking for people who can pay. You see, like "geographical balance," there are also vague percentages on financial aid. Applications with the aid box unchecked and without accompanying scholarship material are often sorted out and evaluated under softer lighting. So don't check the box if they are going to pass this information on. When your acceptance rolls in, immediately write back asking for aid.

You may be worrying about the possibility that all aid and loans will be granted when the acceptances are mailed. This is legitimate. If you are going to use this tactic, you should apply for early admission, so you will not miss the application deadlines for financial aid. Obviously if you truly do need aid to attend the college, acceptance without aid will be an empty triumph. So, before filing, check to see that the early admissions notification date falls before the aid deadline. Don't worry, if they really want you, money will be found. If the school can't come up with something for you, an early admittee, then it's a safe bet that, under normal circumstances (i.e., you filed for regular admissions and asked for aid), your aid application would have been rejected. The point is, you are in—and remember, the university is not the only source of funds. And, don't worry about your acceptance being rescinded because you requested aid after the fact. Acceptances are rarely rescinded on any basis—you almost have to murder someone.

Finally, you can distinguish yourself on the application by declaring your intention to major in something other than pre-law or pre-med. (This does not mean filling out the "Intended Major" blank with Economics and then writing an essay on your dreams of being a lawyer. The last we heard, some of these application essays are actually read and evaluated.) Even if you do want to be a doctor or lawyer, *no* one else has to know that right now. Colleges and universities have to plan their class sizes. They like it when two or three people show some interest in their lonely Serbo-Croatian department. No one is ever going to see your application after the acceptance is mailed, much less force you to concentrate in your "Intended Major." This is a promise.

Now along with the application, you must forward test scores and recommendations. It is said that, in light of the Freedom of Information Act, recommendations are no longer weighed very heavily. It is also said that if you allow your recommendations to remain confidential (i.e., you don't see them) they will be taken more seriously. Whether or not any of this is true, it is in your best interest to get the very best recommendations you can and to get them in on time. So you should check up on the VIPs you ask to write nice things about you, and make sure they remember to mail their songs of praise.

On the other hand, there is no doubt in anyone's

mind that the test scores are weighed very heavily. But the tide seems to be turning slightly in favor of the students, or the testees. Recently, "Truth in Testing" laws have enabled students in many states to obtain copies of the test, answers, and their graded answer sheets. After January 1, 1982, tests, answer keys, and graded answer sheets will be released to *all* students in *every* state. ETS would have you believe it is making this concession because students outside the "truth" states have been at a disadvantage. In reality, when the tests and the answers were made available to the public on this very limited basis, grading errors were uncovered, mistakes in questions were discovered, and, in several cases, controversy arose over "correct" answers. The mind reels when you consider how many errors of all types have gone undetected in thirty years of testing. Obviously, this has raised some question about the tests, and their credibility seems to be eroding. Nevertheless, the test scores are weighed heavily now.

Request a copy of your test—every time you take it! Naturally, you will want to check for grading errors immediately. But that's not all you should do. If you have any doubt that the "correct" ETS answer is in fact the right answer to the question, check it with your parents and teachers. The student who doubted the infamous "Pyramid" answer forced ETS to raise scores all over the country. (This guy should get a medal for academic heroism.) Encourage your friends to request their graded tests. Then you can swap. Every effort should be made to convince your guidance counselor or student counsel to start a file of ETS tests and answer keys. Due to the copyright laws, books of previous tests are not now published and probably won't be in the foreseeable future. ETS may concede the fact that you have a right to study your previous test, but they aren't going to put together a book of old tests for you. Think of all the mistakes you and your buddies might find!

Now about the actual test. It is not as hopeless a situation as you might think. With a little finesse, you can take the test several times and study the answer sheets before you take the test for keeps. In other words, you can practice on the actual test. It's very easy: just never fill in any information about yourself other than a name and "an" address. Use a variation of your name (i.e., nickname or initials) and the address of a friend.

Can you do this? Yes!! ETS is entitled to NO, let's repeat that, NO information about you. Anything you choose to provide beyond a name is used for identification, recordkeeping, and, later, for statistical studies. All of this works against you in some way. If you do provide any identifying information, ongoing records can be kept and your one disastrous score will follow you until your dying day. (While SAT scores are not officially sent to grad schools, they are often written across the top of college transcripts, available to the eyes of one and all forever.) Identifying information enables ETS to record your very first attempt at the tests right above your higher scores of six months later. Both are forwarded to all colleges you apply to. (If you have the stamina to take the Boards four times, ETS will drop the first scores. They will keep only three scores on record for any one student. Isn't that nice?) Remember also that every piece of information you provide about your ethnic/financial/educational background is stored and used by ETS in statistical studies which inevitably prove the validity of their testing methods. (The day ETS allows an outside concern to conduct a study with their data, all the #2 pencils in the world will melt.) Consequently, when you do take the test for keeps, that score will appear all alone, with no earlier catastrophe tarnishing its glow.

When you do take the Boards for the record book, use your proper name and your own address so the colleges can identify you, and know that *you* took the test. But remember, beyond the address, you still DO NOT have to provide ETS with any other information. You should think seriously about it before you do so. You may wonder what all the fuss is about. Well, today ETS has more detailed, cross-referenced information on more Americans than the U.S. government. Supposedly this information is confidential. And it is to a degree— you can't see it, and neither can your friends. If you want to test the bounds of this confidentiality, fill out one of the minority student blanks. Within a month of filing your test application, you will probably get several recruiting letters from schools looking to up their minority enrollments. So much for that. God knows who else has access to this information. When you realize that ETS also makes up tests for the CIA, the implications become frightening. So give it some thought each time you darken a little circle.

Now suppose you did everything and you still weren't accepted. All is not lost. Don't get depressed. After the rejections are mailed out, the school never hears from 99% of the rejectees ever again—they slink off slowly into the sunset. If you really want to go to this school, you should immediately call up the admissions office and ask them why you were rejected. Often merely expressing an interest is enough to win a place on the waiting list. If you can't talk them into that, ask them what you can do to improve your application. The admissions people will rarely say you haven't a chance in hell of getting into the place. They will tell you how to improve the application. If you are persistent, you will get in.

You can go to school any place you want to, and don't let anyone tell you any differently.

1

SURVIVAL 101

Now to the basic facts. You have finally made it into the college of your choice, or whatever. No doubt the orientation week was a gas, but now they are pressuring you to get it all together and register; make your decisions and pick your courses. You just do not know what to do. Have no fear, very few people do. The hardest thing to realize is that this first registration is one of the most important. Careful planning and plotting begins here. Your naiveté can be the kiss of death. Do not be afraid to ask for help. There are people everywhere to help freshmen.

The problem is who to ask. It is a big problem. Do NOT ask the orientation upperclassmen for help. In most cases, these are the rah-rahs, and they will not be ready to get down to the nitty-gritty and tell you about the bads; they see only goods. Everything is good to these people; that is why they were picked for the job— they are the ones who leave the bulk of their estate to their beloved Alma Mater.

Going to the freshman class adviser is also usually a waste of time. Deluged with bewildered freshmen, the class adviser is only going to be able to give you the fast version of tea and sympathy. This can mean big trouble later on. These advisers often know what is going on, but they really do not have the time to give you any more help than explaining the physical process of where to go when.

Your best bet is to see someone in the department that you think you might major in. Do not worry— asking these people for help is in no way considered a commitment on your part. They are usually very willing to help you. But take everything with a grain of salt if the professor who advises you gives you hints on who the good guys are and who the bad guys are. Department faculties are like little kaffeeklatsches; some groups are great friends and others never speak to each other.

Finally, if you cannot get in to see or do not want to go to any of these people, your best bet is your dorm adviser. (Use your discretion—there are off-center dorm advisers, too.) Dorm advisers are upperclassmen or grad students who are there to help you with all kinds of problems, academic and personal. They usually are very nice, and they are not picked on the basis of their school loyalty. They can and will honestly try to steer you away from the horrible courses and give you an idea of what you can expect from the good courses. Also, they are not so inaccessible that you will not be able to beat them to a bloody pulp if they steer you wrong.

In essence, do not be afraid to ask, ask, ask. Careful course planning is the basic key to good grades in college, and you need good advice.

Of course, the advice you will get may be something else again. Unfortunately, most of the administrations and advisers seem to be under the impression that the required freshmen courses are the easiest that you are going to take during your four years at school. THIS IS NOT SO. Required courses follow the peculiar pattern of being beyond everyone's aptitude and totally dull; boredom and the accompanying discomforts seem to be the stuff of which freshman required courses are made. Consequently, you are going to have to force yourself to do most of this freshman work. It is not easy and it is definitely not fun. Freshman year is not a joyride. If you go in expecting it to be one, you are going to have a tough time of it. The shock of reality and the lack of understanding that it will get better is why so many freshmen drop out.

Freshmen have a lot to adjust to. In college, the work goes faster, the classes are generally larger, the grading tougher, and the work load heavier, than what you were used to in high school. On top of the academic readjustment, you have the freedom of your new life-style to contend with. Adjusting to your new surroundings and deciding exactly what is expected of you by your professors are the most difficult tasks you will face as a freshman. Ideally there should be a trial term to allow freshmen to adapt, but in reality, schools just throw you into the thick of it, with a sink-or-swim toss. However, with careful planning, you can get through your freshman year with good grades while you are acclimatizing yourself.

It will help immensely if you know your catalog. Like a good fundamentalist knows the Bible, a good student knows his catalog inside out. Catalogs are not just ads sent out to prospective students, they are cata-logs; they list the available courses and the vital statistics of your school. Your catalog tells you the rules and procedures you will be expected to follow, the departments and majors available, etc. If it is not in the catalog, it does not exist. The catalog is the source of all knowledge about your college. It is your oper-ator's manual—know it and refer to it.

Finally, the grades you receive your first term are very important. High school and college differ in that all your college grades count in graduate-school admis-sions. Some schools will overlook your first year, but these are few and far between. Colleges look at the whole, overall picture of your work and will forgive early mistakes. Not so graduate schools. Erratic grade spans are not looked upon favorably, so even if you straighten out later, these first terms can really drag your average down.

But more importantly, how you do as a freshman is very important psychologically. If you spend the year in misery taking all of your required courses and end up with a 2.0 average, you will be depressed, and not even the prospect of going on to those courses you always wanted to take will relieve it completely. There is only one way to end your freshman year, and that is with high spirits and a high average. Then you will be ready for the hell to come in the next years. Admittedly an easier hell, but hell nonetheless.

REQUIREMENTS: HOW AND WHEN TO TAKE THEM

When you walk into the college of your choice, you are going to be handed yet another catalog and a list of requirements. The requirements are the courses you must take to get the little piece of paper four years later. The list will seem appallingly long, and despite the fact that you will have some choice about courses you will use to fulfill the requirements, you may not be very interested.

Science, language, math, English and humanities are the standard freshman fare. You will have to take one —possibly two—courses in each department. Since you will not have any of the necessary prerequisites, you will end up taking the most general survey course in each department, which is what God and the administration intended. Unfortunately, these surveys are always very heavy on the reading and the paperwork (using about a tree per pupil). Contrary to common sense, survey courses are based on the rather bizarre theory that one can learn the entire history of the world from the moment God said, "Let there be light" to what the president said this morning about our economic recovery being almost complete, all in a few weeks. (Don't laugh—not only do they preach it, but they try to practice what they believe.) Obviously, you should not take only requirements during your first term.

Unfortunately, most schools try to emphasize how important it is for you to get as many of these courses out of the way as possible. The registrar tries to give you the impression that any freshman who does not take as many requirements as he can is going to be far behind his classmates. This is ridiculous. It is more important for you to get the feel of the school and the experience of living on your own. No matter what the registrar and the advisers say, the only course you are *obligated* to take and get out of the way your first term is Freshman Composition. Beyond that you do not have to go.

Aside from the courses you will need for your major, and the basic requirements, every college requires you to take a certain number of courses outside your field. (One must be narrow but expand one's horizons and have a general knowledge of many wondrous things.) This means that sometime during your college career you are going to have to take a few courses you are vaguely interested in, though they are not connected with your specialty. What better time than in freshman year? Even the gut that you discover in the psychology department will count toward your degree as one of these outside or remote hours. (Unless you major in psychology, in which case it will probably count as a major elective.) Do not let anyone fool you into thinking that every course you take your first term is not totaling up points toward Diploma Day.

Aside from saving yourself from a life of boredom interspersed with hours of sheer drudgery, there are definite advantages to spreading your requirements over four years. When you do get around to taking that science requirement in your junior year, you will be an old hand in a class of mostly inexperienced freshmen.

Not only that, but you will have friends who suffered through the course before, who will be able to tell you about the various professors available and whom to request—to say nothing of helping you with notes, tests, and papers. While these courses might have seemed incredibly tough to you as a freshman, later you will recognize them as just basic survey courses and they will be a cinch (because you will have learned how to deal effectively with a survey course).

There is also the thought that by the time you are a junior you might have taken one or two of the prerequisites to get into something other than Western Civilization I. Then you might be able to fulfill the humanities requirement with something more stimulating than a sixteen-week course on the history of man. Look around and do some thinking. You might not want to take Geophysics 307, but the Influence of Television on Modern Society is a hell of a lot more interesting than History 101 or Sociology 101.

When you go in to register for your first term, take your Freshman Composition and one other easy requirement. To make everything even easier for yourself, start your language requirement now, too. (More about that later.) Look over the catalog. There are lots of things in there that interest you. (Even if there are not, talk yourself into it.) Take something you like. If you take a course that you are interested in, it will make the adjustment to college much easier, because you will not have to force yourself to work at it. And then, to round out your program, look for a gut. There are guts on every campus. Do some research. Ask upperclassmen. When you find one, take it. You are going to need every minute you can get during your first term, and the easier your courses are, the better off you will be. A smart freshman's program might look like this:

English 101
German 101 (or whatever)
History of Photography (your interest course)
Italian Renaissance Painting (campus gut)
Philosophy 101 (a distribution requirement)

As we said before, you are going to feel a lot more confident coming back for your second term if you have done well the first time around. To the outside world, a 3.5 is a 3.5; no one has to know that you got it with History of Photography or Italian Renaissance Painting.

FRESHMAN COMPOSITION

Every school requires it. There is almost no way out of it. When you come to register, the English department will evaluate your particular level of illiteracy on the basis of your SATs, placement tests, and high school English grades. The only people who are excused from this great college institution are in the vocational trade colleges or in the Earth Moving Equipment School in Texas (since they obviously have no great need to be able to express themselves in eloquently written prose). Resign yourself.

If you are deemed abysmally illiterate, you will have to take the standard Freshman Composition. But this course is absolutely nothing to worry about if you can express yourself in sixth-grade-level English for the weekly paper, and if you can read Monarch plot summaries for the class discussions. You must take care, however, not to be a dead body in this class. Some of the larger state universities use this course to cut down on the number of their freshmen. Witness the presence of the notorious Mr. Flunkin' Duncan who wandered the Ohio State University system in the early seventies. There are gas station attendants all over the Midwest who can tell you about him. So ask a question or two, take the tests, and hand in the papers.

For the semiliterate, there is advanced Freshman Composition. If you get placed in this section, it is great. There is generally a wide choice here, and you can pick some specialty such as a course on Women in Literature, Cinematography, or the History of Modern Literature. Beware, though: these classes are smaller than the regular freshman comp classes; your absence will be noticed and sometimes noted. As always, with the disadvantages there are advantages. While your papers will have to come up to a higher standard and be somewhat longer than those required in Composition 101, you will not have to write a paper every week. Do not worry about this being a course for freshman English majors who will be stiff competition. English majors who have not already declared their intentions and demonstrated their high degree of literacy have yet to surface. English, like sociology and psychology, is a major that slowly gathers the people who have become disenchanted with their chosen fields. They are the seeds that fall to the wayside. Any potential English majors you meet in these classes are great people who tend to be a lot of fun. Get to know them. They are usually the ones who throw the best parties in their junior and senior years.

For the chosen few who are considered literate (usually on the basis of advance placement English exams), the English requirement is waived. If you are one of these lucky people, you can throw away your Monarch notes to *Hamlet* and *A Tale of Two Cities* with a clear conscience. Obviously you are a master of the written and spoken word. But do not think they are going to let you off the proverbial hook so easily. If you do major in something other than English, be prepared to be nagged constantly by your major adviser in years to come. He or she will review your transcript and say, "Isn't this program a little one-sided? You never read any books. How are you ever going to talk to anyone? Or write a decent letter?" Aha, the secret is out. So that is what these courses are all about!

THE LANGUAGE REQUIREMENT

The language requirement is a big drag and is almost impossible to get out of unless you were raised bilingually and can prove yourself fluent both written and orally in both languages. Some schools may exempt you on the basis of your high school record and language achievement but these are few and far between. Before they exempt you, they sometimes require you to take their placement test. If you score next to native perfection on this one, too, then you are out. But not before. Even if you should prove yourself, they will usually "very strongly recommend" that you take a literature course in the language you are so fluent in.

If you opt to take their placement test, be aware that it is a double-edged sword. If you do not score high enough to be exempted, they will place you in the semester level that your score merits. If you cram, thinking that this will pull you over the necessary exemptions score and then do not make it, you are going to be left high and dry in a level that is beyond you. Everyone knows that crammed knowledge seldom hangs in the mind more than twenty-four hours. So beware. You will have to do mucho explaining to drop back a level. But if you feel you have a chance to place out of the language requirement, try by all means. Try at the first opportunity that the test is offered to you while your high school or traveling knowledge is still with you.

If you took French or Spanish in high school, hated every minute of it, and cannot stand the idea of fulfilling your requirement with another word of it, you can always start a new language in college. One of the better aspects of college is that your realm of choice is greatly expanded. If you have always had the dream of learning Polish so you can speak to your grandparents, or Italian so you can curse out your neighbors, or Portuguese so you can run away to Brazil, or Chinese so that when you become famous you can defect, college is the place to learn it. You may have to start a new language anyway. Many majors strongly recommend that you have a background in a specific foreign tongue. (Italian for art history, Russian or German for engineering, French for literature, etc.) If you are going to start a new language and have no driving passion to pursue, look into what language is recommended by your prospective major department. Now while college language courses are not like your high school French class plodding along, they are not the Berlitz school either. You have nothing to fear. The first semester of any language is very easy. While high school was the "Sesame Street" version, college is at the "Electric Company" level. If you keep up with the daily work, and study for the little vocabulary tests, you cannot miss getting at least a "B."

Starting a new language can be a drag because you will probably have to take two full years of it, but the beautiful grades you will get if you do keep up with it are nothing to sneeze at. The recommended languages (for ease and comfort) are Italian, Spanish, and German. Here you can just whiz through, even if you are a linguistic clod. By the time you finish the two years (any masochist is welcome to the full four), you will be able to get along pretty well in the chosen tongue.

It is the one concrete skill you can pick up in the liberal arts field.

THE SCIENCE REQUIREMENT

Fulfilling your science requirement is an even bigger drag (it seems to be the only appropriate word) than fulfilling the language requirement. Here there are no placement tests, no high school credits given—in short, no way out. No matter who you are, or what you are going to major in, you have to take it. It may appear to be an insurmountable obstacle to the budding poet or diplomat or lawyer, but it is not as bad as it seems. In almost every school across this wondrous nation, you will find that there is a science for the nonscientific minds among the student body. This course is not-so-pure but simple. It is the one most people use to fuffill the requirement. It will be easy. In some schools it is geology (rocks for jocks), in others astronomy, and in some physics (physics for poets). Often it can be psychology. The trick is to get your feet on the ground at your college before plunging into the science requirement. If you jump right in before investigating, you may be dismayed to find that your course is a bitch really meant for science majors, and that there was an easier alternative.

Fulfilling the science requirement is really a matter of preference. Be aware, however, that even after you track down "the science gut" among the various options, you are going to have to work a bit harder at it than at your other courses. If you feel that you are up to taking chemistry or astrophysics, by all means try it. Just keep in mind that these are rarely the easy poetic sciences. While you may have a great time pretending to be Dr. Jekyll during the term, you should be prepared to take on the role of a martyr on the eve of the final. The only thing that will save the day for you is remembering that nonscience majors are usually permitted to take the science requirement pass/fail. Think about it. Is your fantasy fulfillment worth a "D"? If not, take the course pass/fail or take the gut.

Finally, if you are going to take the gut science, make sure you have looked into it. If psychology is the way, check it out to see if it takes two or three terms to fulfill the requirement. Since psychology is not one of the hardcore sciences (and the first term tends not to have a laboratory section), many colleges require that you take an extra term in order to have two terms of laboratory work. Also, many schools have two different sections to begin each science, one for the poets and one for potential majors. MAKE SURE YOU ARE IN THE RIGHT SECTION. The major section is going to be considerably tougher. Remember, if you are even remotely considering this science for your major, the nonmajor section will do you no good (since they will not count it for your major). If you suspect that you are going to major in this science, wait until you are damned sure before taking the basic science course. It is no fun wasting your time retaking a basic course. And you will probably have to retake it because it will be the prerequisite for the more advanced classes and seminars later on.

At the risk of editorializing, it has to be said that the science requirement is a waste of time for all the people involved. The poets could not care less; if anything the apprehension involved heightens their aversion to anything scientific. The science majors are not fond of spending time in large classes packed with those other warm bodies who merely slow down the learning process. Without the poets taking up the seats and the professors' time, the majors would be able to have more individualized instruction and the class could cover much more territory. But that's the way it is. The college thinks of you as flaming volcanic lava, and it wants to stuff everything possible into you before you cool and solidify.

PICKING YOUR PROFESSORS

No matter what anyone says, the professor and his method of teaching is what makes or breaks a course —and you. A professor can teach brilliantly and make the dullest books seem like the bright spots of your week. Another can attempt to teach a course on ZAP Comics and make it sheer agony. So you can see that not only do you have to be careful about picking your courses, you have to look over the available teaching talent as well.

Professors in General

You can find out a lot about professors simply by asking people, and we do not mean your advisers. (They do not count as people; they are part of THEM.) They would never tell you, never, ever, ever. There is honor among thieves—and the same goes for professors. (You would not rat on a fellow student— we should hope not, at least—so how can you expect professors to criticize their own?) Ask people in the class ahead of you. What have they heard about Professor So-and-So? They will know. If it is good or bad, file it away for further reference. If they do not know anything, then it is chancy. In college, the devil you do not know is *never* better than the one you do. You would not dream of putting down a heavy bet without a hot tip on the jockey—do not do it here, either. Look for someone who has taken a course from the professor (preferably the same course, since some professors switch personalities with each course). Find out where you stand. If the person still has his papers and class notes, then you will be in a good position to judge the course and the professor. If he is a widely beloved "A" giver, walk into his class with a smile. If no one has ever heard of his giving an "A" to *anyone,* run in the other direction.

Course evaluation books will sometimes give you a hint as to a professor's talents, but do not depend on them. Every professor has his groupies, and they are the ones who stay behind class long enough to fill out the questionnaires. The rest of the class may have left in disgust, or if they did stay, their comments may have wound up on the cutting-room floor. But more about that in the next chapter.

So ask around—and ask and ask. Find out who actually teaches the course. In the catalog it may say Professor Famous, but in the hard light of day, it is a graduate student who trots his pompous ass in twice a week to read the old lectures. (They were good enough for the class of 1912 and they are good enough for you, jokes and all.) Graduate students are to be avoided like the plague. They grade tougher, are not crazy about discussions (you might bring up something they are not too strong on by themselves), and they are deadly serious (a rattlesnake has a better sense of humor than a graduate student assisting Professor Famous). Impressing the professor is their primary concern, not you. Do not forget it.

Professor Types

There are a few professor types that are commonly found in and around most campuses. They are the same on almost every campus, from Dartmouth to Slippery Rock State U. (Only the names are changed to confuse the innocent.) As time goes by, you will learn which ones your personality and talents please most, and which type you can and cannot stand. They are not always black and white, however, and we should go into a few specifics before strolling along:

Prof. Cool: This is the guy who still wears love beads and black turtleneck sweaters. He spans the generations of the beats and the hippies, although he should have died with the sixties. His lectures tend to be long, irrelevant discussions of why you are in college at all, interspersed with diatribes against the system. If it is a small class, he specializes in encounter sessions; even if it is an engineering class, this professor is going to try to relate the material to YOU, very personally. Not only is this embarrassing, it gets to be a bore. (He cannot get it into his thick mind that you are paying mucho for this hour, and that you want information not games.) If you take a course from this type of professor, keep in mind that the class is going to be called off in sympathy with every strike in the nation, every natural disaster in the world, and in memory of everyone who has died since Adam. God help you if there should be a strike on campus. You should be prepared to run down every human value (except, of course, love). The clincher is that when this professor gives a test, it is a bitch, and he grades it like a little Hitler— because HE HAS TO BE HONEST.

The Professional Sympathizer: This is the professor who is young, if not in body, then in spirit, and still remembers his own undergraduate days fondly.

(He has delusions of grandeur, thinking himself a mixture of Ann Landers and Mr. Chips.) He tries and bends over backward to make the material interesting and the tests reasonable. While he sounds like a dream, he can get you into big, big trouble. If you cannot get your paper together, he gives you an extension with no fuss. He puts off tests if everyone else is giving tests that week. He lets people take incompletes as if they were going out of style. He forgives and lets everyone get away with everything (he gives you enough rope to hang yourself three times)—and he forgets. And we do not mean he forgets simply to buy laces for his laceless shoes; no, he forgets BIG. This is the guy who will get your grade to the registrar about a year after you took the course, if you are lucky (and have been constantly reminding him). Sometimes he hands it in the day before graduation; then again, he may not hand it in until six months after you were supposed to graduate. While basically this professor is a nice guy, he belongs in an Elks lodge, in a small-town library, in an automated lighthouse—anywhere but in a classroom.

The Professor Who Can Learn from Every Student: Someone should point out to this fool that you are working evenings to help pay for your courses and that you expect a professor to at least lecture for the money. This is the type who does not lecture, but proceeds on the assumption that everyone has something of value to add to the class's understanding. (It never occurs to him that most of his students are there to learn, not to teach.) Unfortunately, there are those who have a lot to say, but there are not many who have anything to say worth hearing. Then again, not everyone has anything to say, or wants to say anything, but they are encouraged to speak up and "contribute" anyway. The classes end up spending hours listening to one or two loudmouths in the front row. It is infuriating and boring. And the professor is a fourteen karat phony. Although everyone has a valuable opinion, something happens when this "valuable opinion" is transferred in writing into a bluebook. This professor grades like a hanging judge. Some pigs are more equal than other pigs.

The Dedicated Scholar: This professor does not live in the real world. He is easily recognized. When it is raining, he leaves puddles behind him because he did not realize it was wet out there. Nothing will come out of his mouth that does not directly relate to the subject. If you take a course from this type of professor, understand that this is his raison d'être. It is not just a passion; it is his ruling obsession. He is doing you a favor by teaching you all these wonders, and he genuinely wants to interest you in the subject (although if you are not he is totally baffled—you mean you *don't care* why Milton wrote *Paradise Lost?!*). Do not take him or his course lightly, and never joke in class. When taking a test, realize that no new ideas are appreciated. The professor taught you the correct theories and facts, and he expects a direct feedback. If you can take this type of dedication seriously, this professor is your speed. The tests are nothing more than a review of class lectures, and you are never expected to make a contribution in class other than yes, or a thoughtful nod of understanding. Needless to say, this professor will not cause you great mental strain, much less stress.

The Prehistoric Man: Believe it or not, universities harbor people who have not read a new book on their subject or rewritten a lecture since the dawn of their tenure (some thirty-five years ago). This is the type who will lecture from old notes (taken by him from his old professor who was born around the same time as Jesus Christ). They usually have graduate students who grade the papers for them. The tests are usually the same as those he gave to Henry Kissinger in his freshman year. It is very easy to take a course from this fellow if you can find someone with a notebook and copies of the tests from a previous term. You do not have to worry about this professor at all—you will see him only in class, his office hours being nonexistent. He does not expect or want any visits or questions. He lives a vacuumlike existence. Anything you are going to learn from this type is so old that it is useless, or it is dead (if not dust). This type of professor is very reliable; like Old Faithful, he gushes forth the same thing at the precise moment calculated. But he is hardly what you would call stimulating.

The Firebrand: This profesor is most often found in the liberal arts area teaching Black Literature, Woman's Literature, Contemporary Civilizations, or Sociology. Occasionally you will find him teaching political science (Modern Revolutionary Politics). If you can stand the heat, jump into the fire. This professor will expect a lot and demand much in terms of work. But he will also teach you something other than the subject matter at hand—he will teach you how to think. If you do not mind hearing someone really lay it on the line and tell it how it really is, etc., join the throng trying to get through the door. You will learn a lot about the subject and a lot about yourself.

The Real-World Professor: Here we have the best of all possible worlds. This is the professor who teaches because it is his job. He is dedicated but does not let his knowledge stifle an enthusiastic undergraduate. He is nice and very sympathetic (but firm—he does not let his students beat him into submission). He is enthusiastic but understands that you are taking other courses and does not try to load you down with reading for the other six days of the week. He grades the papers himself and sees people outside of class hours. But most important, he knows that there is another world outside the university. This is the professor to see when you are having trouble keeping your problems in perspective. He is so calm and normal that all your cares will fly out the window. He is the type who can say to the rejected premed, "You do not have to be a doctor. It wasn't stamped on your forehead at birth." He can tell the flunkee, "You can always leave town and change your name," and mean it. If he asks you over for dinner, go. This professor is not trying to impress you or his colleagues; he really likes you. The food will be good.

There they are. Of course we could not give you a rundown on everyone, but these are the basic types. In every category there are subdivisions and weird exceptions. Of course, you rarely run into pure types. And there are some professors lurking about who are deadly combinations of two or more types. You just have to check around and feel them out. Each professor is going to have the fate of your academic career in his hot little hands, so do check him out. You would not let the little man with the wild hair and blazing eyes drive you home over an icy mountain road, would you? It is the same here. Your life is your life. Enough said.

THE VALUE OF THE COURSE EVALUATION BOOK

Many people swear that the campus course evaluation book is the only way to pick a course or a professor. Relying on a friend's opinion is not as good as the distilled statistical knowledge you can get from the course evaluation book. (These same people believe *Rolling Stone* is a grassroots underground paper.) Ten years ago, when course evaluation books first appeared on college campuses, this was so. At that time they were considered tremendous innovations, and appropriately so. They created an understandable furor among the faculties. (Students judging their professors and giving them grades was not only preposterous but blasphemous!) However, as time has gone by, the course evaluation book has become part of the establishment (if you cannot beat it, absorb it). Many universities subsidize the printing of the books through the student activity fund, and of course, by now you all know the power "funding" carries. Many of the staffs have become filled with the more diligent and serious-minded students who are filled with the appropriate awe of professors and who take any critical comments toward the revered faculty (you know, those people who live with the gods on Mount Olympus) as personal insults. These people gently edit adverse remarks, casting aside anything stronger than "although Professor Supreme is comfortably rumpled in appearance . . ." Such cute criticisms are considered strong enough to show a balanced picture of student opinion. Finally, the kiss of death for the course evaluation movement was the assignment of faculty advisers. Needless to say, the view you now get of the faculty can be slightly biased, if not warped.

The most useful part of the course evaluation book is a point of debate. Some people rely upon the student comments. Many people insist that they are the most deceiving. Student comments can be edited—and brother, are they edited. One Miami University sophomore we know watched a faculty adviser cut her copy to ribbons and insert glowing reviews by ficticious students. You must also keep in mind that every professor, no matter what kind of bastard he is, has his fans who write voluminous wonderful comments. They are the ones who are willing to stay behind to gush for the cause. People who tend to write a more perceptive analysis than "the most fascinating course I ever took" find that their comments hit the floor, being considered "dull copy" by the evaluation staff. The student comments that you should rely on are the ones you hear in the dorms and in the dining halls (straight from the horse's mouth), not the ones you read in the guide.

Another bad thing about the guides is that most schools give the professors the option of having their course evaluated, and many other professors and their courses are simply overlooked. Your guide is not comprehensive and it is not objective.

This is not to say that the course evaluation book on your campus is not worth looking into and buying. It simply means you should not take the information as the be all and the end all, it is not the gospel truth. Salt well.

The useful part of the book, and what you should really pay attention to, is the grade span index printed in the front or back of the book. If your book does not print this, then do not bother buying it. If it does, by all means pick up a copy.

The grade span index will give the range of a particular professor's grades. This way you will be able to pinpoint the people who give "Cs" for the good work, "Bs" for the exceptional and save their "As" for the genius who surfaces with every blue moon. These are not the people to look for, because even if you do well you will get a mediocre grade for your trouble. If you are convinced that you are going to get that one "A" that the professor is going to give this year, then take the chance; if you know that that is not going to be you, look further. The best professor is the one who gives an average grade of "B"; his grade span will look like this:

Professor Lovely: A B C D E
                  35 72 29 1 2

Now in that class you have a chance to be one of the 107 people out of 139 who came away happy. Unless you died in your seat there, you could not miss.

The people to avoid look like this:

Professor Nasty: A B C D F
                  2 4 13 4 1

Many schools do not even attempt to put together a course guide. Actually, the best way to judge a professor and his grading method in a course is to ask around. It may be the old way of doing things, but it is really the best. If the course is really bad (and will quite literally bore you into an early grave), everyone on campus will know it. Ask your friends, and be specific with your questions. Do not be afraid to ask what you want to know. Can I get an "A" with a minimum of work?

Finally, if the professor you have in mind is totally unknown and you feel you want to give it a try, do. You can always add a class during the first weeks of classes. Do a bit of research before you decide. Take a hike to the bookstore. It will have the reading list on file and will be able to tell you what material is going to be covered. You can check out how long the list is, and how many of them you have read. If you do not read much or very quickly, you can see how well known the authors are, and if it will be easy for you to get notes and criticisms on the required books.

Course evaluation books can really leave you up the creek if you rely on them. If you use them correctly and supplement their factual knowledge with the comments of your fellow toilers and a look at the bookstore lists, you should be in good shape and know what to expect. But if you just use the guide, you are in for some surprises. All that glitters in these books is definitely not gold—just touched-up tin with sharp edges.

READING LISTS AND HOW TO JUDGE THEM

Reading lists can be very, very deceptive. No professor passes out a list that will be completed or adhered to strictly. Remember that there is a sort of perverse honor won when a professor hangs up the longest list on the department bulletin board. It is one of the trappings of intense scholarship. It should not depress you or dissuade you from taking a course. It is an unwritten rule that the longer the list is, the less the students will actually have to read.

If you are at all apprehensive about reading and trying to keep up with speedy classmates, then go to the bookstore that will be stocking the books for the course. Look at the lists for the courses you are thinking of taking. The list that the professor gives the bookstore is the one that you will be expected to buy and to actually read (and even this is probably a little overestimated). Invariably, this list will be shorter than the one on the department bulletin board and the one passed out on the first day of class. Aside from being buoyed by the shorter number of books, you can check to see how many of them you have already read. (This is when you appreciate the fanatic you had for a teacher in high school.) This can be a big help, especially if you are not an avid reader. Your reading speed will probably pick up as you learn how to skim material in college (and you do learn to skim, for it becomes sink or skim—terrible, yes, but we try hard). However, in the beginning, it is good to begin on ground that you have already explored somewhat. It is less likely to open up and swallow you unexpectedly.

Do not lose your head at the bookstore and buy all the books for your courses. You will be wasting your money, and you will end up with a bookcase full of untouched books. Pick up the basic textbooks for every class you plan to take as quickly as possible. These tend to sell out in minutes. (The cynical bookstore operators never believe the estimates the professors give them—mainly because of all the books at the end of the list which never sell.) And of course the reserve libraries never have enough books to accommodate the unlucky. But other than the basic texts, buy the rest slowly. In most cases, by the middle of the term, you will discover that unless your class covers one book per session, there is no way they are going to make it to within calling distance of the end of the list.

Finally, if you are one of those people who picks a course on the basis of the length of the reading list, remember what you actually have to read to get a good grade is even smaller than the list of what you are "required" to buy. In any English or literature course, a smattering of knowledge will do. You never have to read more than the plot outlines and author's themes of the books involved to write an essay question. (Objective tests are very rare in reading courses.) For history courses, you need only a good basic text dealing with the time period you are studying. All the shadings and historical or political theories you will need will be provided in class. Just listen and take notes. The same holds true for political science, philosophy, psychology, etc. If you attend class regularly, you do not have to read much of anything. Get with it and save your money. Besides, you would have to spend every waking minute reading if you read every book on every reading list pased out to you in the course of your four years, and that is not including the extra recommended reading for those interested in going into the subjects in depth.

ART, CINEMA, PHOTOGRAPHY, PHYSICAL EDUCATION, AND BASKET WEAVING

Now that you have determined what "hard" courses and material you are going to take this term, you have to find a pleasing gut. Colleges abound in gut courses. If you look around carefully and listen to the upperclassmen, you find not one but scores of them exist. You have to be careful, though; sometimes the course will sound like great fun in the catalog, and you find out

later to your sorrow that it is a zinger. (Also check to
see that the course carries full credit, sometimes they
only get you a partial credit.) It is not just the freshman who can get screwed in this manner, it has been
known to happen to the most worldly seniors. A happy
group of last-term seniors at Columbia decided to take
a gut course together as a final fling. One of their most
respected wheeler-dealers picked the course and everyone signed up in high spirits. History of Television (the
name of the fatal course) should have been sixteen
weeks of Lucy reruns. It was not, and the group failed
en masse. (Who was to know it was going to be a study
of the political machinery of the fifties?) So watch out.
Make sure it really is a gut.

Most guts are in the arts and related fields. You can
determine this by asking a few friends if they have
heard of the course. Every campus has one or two
courses that are not just known but are notorious.
Nearly everyone since time immemorial has taken the
course at one time or another. You can get picky and
look for a gut that is not only easy but fun. They are
not hard to find; you just have to ask around. For the
sake of brevity, we will go through some of the departments that seem to specialize in giving gut courses.

Art: Actual art courses can be pretty tough. They are
fun, but very demanding and time consuming. You
will have to get to work very early on your required masterpieces because it is much harder to
come up with instant masterpieces than instant
papers. Also, your work is graded on talent, which
only goes to make it that much tougher if you are
short on talent. (Some of the teachers would give
Leonardo a "C.") If you cannot draw, do not
take any practical art courses. But if you want to
try it, do not take Painting I or Drawing I. Hidden

in the catalog somewhere is a course in welding—
that is, metallic art. It is a gas. You can come up
with anything because here the theories of modern
art prevail. Materials, however, are very expensive.
And you have to lug your masterpiece away after
it has been graded.

Better than the practical art courses are the
theory of art, or art history courses. Art history
is one of your best bets. Many people will scare
you away with the warning that in art history, the
tests will be slide shows where you will have to
identify paintings studied during the term and a
few unknown, unseen paintings by the artists
studied (to see if you can recognize style). This is
nothing to worry about. When the lights go down,
the hum of consultation arises. If you are going to
miss the mystery paintings (or any of the paintings, for that matter), so will the rest of the class.

Cinema: Cinema courses are gut *par excellence.* They
do not require you to make a film—you can write
a critical paper, write a screenplay, or adapt a
novel. It is almost always up to you. The history of
film courses are the best. As for the tests—they
actually give tests which make the quizzes on the
back of the Wheaties box look like GREs. Film
cannot be beat. They show movies during class
so you never have to force yourself to go. And if
you take a contemporary film course, you get to
see recent movies for nothing. Enjoy!

Photography: Photography, on the other hand, is one of
those courses that sounds great and then turns out
to be an absolute bitch. Not only is it terribly expensive (you must provide your own camera, etc.,
and Polaroids are distinctly frowned upon if not

banned), it is also time consuming. They do not let you send your film to Kodak in Rochester to be developed. Oh, no—you must do it yourself with your own two little hands. You also have to learn all about cameras, lens angles, all the mathematics involved, how to develop the pictures, and last but not least, the theories of picture composition. Your photographs of "Sunset Over Cincinnati" will not be appreciated.

Basket Weaving: This is the nice name for the easy courses that abound in almost any department. Even the physics department will have one or two. They are courses designed to introduce the one-sided science or math major to the world outside his narrow little field. If you can carry on a normal conversation with another human being, you have got it head and shoulders over the other people who will be taking this course. While it will be tough for them, it will be a gut for you. Look carefully through your catalog. In chemistry it is likely to be called "Chemistry and Its Influence on Modern Society"; in architecture, it will be "The Architect in Society"; in math, "The History of Mathematics." In business it will probably be called something like "The Concepts of Office Management"; in psychology, "The Development of the Community." Get the drift? It is always the course that is most general in concept and the vaguest in actual content. Find it and you have hit the jackpot.

Now that you know what a gut is and where it can be found, you should have no trouble tracking one down to suit your tastes. The only problem is getting into it—because you usually have to climb over people to get in the door. But even then your school will not let you down. To keep you in shape for climbing over bodies, running to the registrar, and simply maintaining the stamina to take all of this, your school gives physical education courses. Most schools require a minimal number. If your school requires them, you are going to have to take them.

You see, the blind can take swimming. The dumb can do anything except cheerlead. The deaf will be excused from folk dance. In essence, unless you are a quadriplegic, you are going to have phys. ed. if it is a requirement. There is no way out. Do not kid yourself and think that they are going to be sympathetic to the pleas of a senior with completed academics and no phys. ed. Regardless of your 4.0 average and your Phi Beta Kappa key, they are not going to let you stand on that graduation line unless you have proven (by surviving so many terms of phys. ed.) that you are physically fit enough to stand. Gym, however, does not have to be all blood, sweat and tears. Here are some courses to consider, and a few to avoid like a swarm of locusts.

Badminton: Ah, the beauty of the shuttlecock as it crosses the net. It is the game of gentlemen and guaranteed to be a breeze. The only problem: it is boring—very, very, boring.

Archery: Archery is better. You walk between the target and the shooting line. You shoot, you walk, you fantasize about being Robin Hood or William Tell or simply about defending the castle from barbarians. It is guaranteed not to raise a sweat. And it does take skill and concentration, so you will not be bored.

Swimming: It is horrible in the instruction classes. But if you can get into one of the free exercise classes, you have got it made. Go to the gym and suit up in one of the nifty little red (why is it they are always red?) regulation suits. Go out, dive in. With thousands of red-suited little bodies swimming up and down the lanes, the instructor is not going to know whether you have done one lap or the required ten or twenty or whatever. So swim up and down a couple of times (you do not have to stay until you feel as if you are about to drown) and leave. The only problem with this type of course is that the phys. ed. department will allow you a liberal makeup schedule (another case of giving you enough rope to hang yourself). Do NOT let this get away from you. Suddenly you will find yourself with a couple of hundred laps to do, and only two days to do them in. Aside from the very real danger of drowning midway, there is the problem of getting them all done. When there is only one little red-suited body to watch (and it is crawling back and forth at an agonizing slow pace) it is easy to count every single lap.

Tap Dance: A definite gas, if your school gives it. No one in the class takes it seriously. The girls stand around pretending to be Ann Miller and the guys think they are George M. Cohan. Buying shoes can be expensive, and without them, why bother? If you do not have the shoes, you are going to miss out on much of the fantasy. And when you get drunk at parties, you will not be able to strut your stuff properly.

Basketball: Basketball is for masochists. People in this class are actually trying to get into shape, and they are not into fooling around. It also seems to attract the more vicious and aggresive types. Take this and you will need the rest of the week to recover.

Folk Dancing: It sounds like fun, you say. It isn't. Sweaty palms abound.

Fencing: Fencing is a definite gas in the beginning classes. The intermediates tend to take themselves too seriously. But if you want to see what great fakers Errol Flynn and all the others were with the old buckled swash, take it. It is medium strenuous.

Yoga and Exercise Classes: These classes are taught by very serious people. They do not laugh, and you'd better not either. For pretzels only.

Crew: Crew is really dreadful. Sometimes the class will practice on a river or a lake, but, for the most part, you hit the water just before the race. The rest of the time you practice in a basement tank. It is very strenuous, very smelly, and NO FUN.

Tennis and Jogging: These are only for the serious. No fun seekers here.

Bowling: You think you are going to have fun here, don't you? You aren't. Most schools do not have automatic pinsetting machines in their alleys. Can you guess who sets up the pins? Right. You spend half the period hoping to knock the pins down and the other half secretly praying that none knock you down. It is the pits of physical education.

There you have it. You have to be careful picking a physical education course. The wrong one can be fatal or very nearly so. If you can admit you are a meatball, you can always take ping-pong. (Good for future diplomats.)

LARGE LECTURES

After you have picked what you want to take, you have one final option; whether to take a lecture class or a seminar. Much has been said and written about the large lecture classes on college campuses these days, and most of it has been bad. But do not let that sway you. There are definite advantages to a large lecture class. There are admittedly also disadvantages; but for a freshman, the advantages outweigh the disadvantages.

In a lecture, you will have friends to study and compare notes with. If you do get behind in the work or the reading, there is a minimal chance that you will be called upon to display your ignorance to the class and the professor. The tests given to so large a class are not usually very difficult. Long essay exams are seldom given since the professor would have to set aside weeks to grade all the papers. (Yes, we know. One of the disadvantages of a large lecture class is that often the professor does not grade the exams and papers himself but has a graduate student to grade them.) Finally, for a freshman (and everyone else for that matter), a lecture class is one of the best places on campus, aside from the dormitories, to meet people.

The problem with the lecture class is simply its size. In many schools, the survey lecture classes have grown so large that the administration had to install closed-circuit television for those in the back reaches of the auditorium. Needless to say, if you feel that you are one of those who is easily lost in a crowd, this situation is not for you. There are other problems that go hand in hand with this limitless size. If the professor encourages comments, one or two loudmouths can take over very effectively. In many cases, there is almost no opportunity to ask a question. If the professor loses you halfway through the hour, you are going to stay confused unless one of your friends was paying closer attention. If you try to get in to see the professor during his office hours to clear up the point, you may have to wait for an hour or two to get five minutes of his scarce and valuable time. To put it bluntly, you will be part of the teeming multitudes in a lecture class—expect to be treated as such.

Despite the disadvantages, this sort of class is really the best for the freshman (and the most fun for everyone else). If you have half a brain, you can get through the exams. You will have lots of company to call on for help if you think you need it. (Also, you will not have to go to class if you do not feel like it. You will not be missed and there are many sources to get the missed notes from.) It is a good example of community cooperation and effort. It is a very social gathering during the classes and during the tests. You will come to understand the saying that adversity ties people together.

SEMINARS

Seminars, on the other hand, are very small classes. The maximum size of a seminar is usually under fifteen people. (The professor's dog does not count.) Many times it is smaller than that. The seminar is never a survey course. It is an intense and exhaustive study of one aspect of a particular subject. Do not kid yourself: a seminar is not a picnic.

The reading lists are very long, and you will be called

upon—by name. With only ten students in the class, the professor is going to get to know you very well. He will always know exactly how well you are doing, and he will be a good judge of your dedication or lack thereof. Aside from these perils, there are the tests, which tend to be long and involved essay exams. A seminar usually entails a lengthy paper as well. If that is not enough, you should understand that attendance is very important. The class is small enough for the professor to note your absence. He will ask what happened to you last week (usually with genuine concern) when you do show for the next class. Finally, you can never use anyone's old notes for a seminar class. A seminar entails lots of discussion, and discussions move in the class's direction. It is tough, as you can see. But it can be the most rewarding course you take, since it is usually in seminars where you learn the most (through no fault of your own).

Obviously, these classes were not designed for the underclassmen (freshmen or sophomores). They are meant for upperclassmen (juniors and seniors), who can cope with the trials that they impose. This does not mean that there are not advantages to seminars. There are. You get individual attention. You learn a great deal. And you get to know someone on the faculty well—well enough to ask him for a recommendation at the end of the term. You will be entitled to spend time with the professor during his office hours; sometimes he will even set aside time for lengthy consultations. Underclassmen, by the way, often have to get special permission to take seminars. They are a good method of instruction for those who are ready for it. Do not try them until you are ready.

LECTURES VS. SEMINARS

When you get down to it, these are the two types of classes that will be available to you: the lecture and the seminar. No one but you will be able to decide which type will be your cup of tea. It depends on how you want to balance your schedule. You can see that they both have advantages and disadvantages.

As an underclassman, it is best for you to look for a course that is somewhere between the two. If you make an effort and look around, you can find courses that have enrollments of thirty to forty people. These classes are a happy medium. You may get called upon every once in a great while, and yet the professor will not be aware of your name. You will be able to get in and see the professor during his office hours for more than five minutes; yet he will not have time to focus on you and your problems for an hour or so. If you get into a seminar, some professors like to see their students individually and even set aside an hour or so every week for each and every one.

If you do get into a seminar right off the bat, be aware of the pitfalls we have outlined and be prepared for some stiff competition. There will not be many other underclassmen in the class, and you may simply find yourself out of your depth. If you find that the parade has passed you by, do not suffer in silence; that is what the professor's office hours are for. As a seminar student you are entitled to a bit of his time and expertise. As an underclassman, though, these courses are just not for you.

The lecture, designed as a survey of whatever subject, is much more your speed. You may never see the professor alone, and you may have all your papers graded by a graduate student, but you will be able to keep up. If you have trouble in keeping with the class (or getting there), there will be plenty of people around to ask for help. Because of the lax atmosphere of the lecture, and the fact that no one is going to know whether or not you are doing the work, you have to watch yourself and make sure you do something. You can get lost very easily. You have to make sure that you do not get too far behind, or that you rely too heavily on others to help you out. Otherwise you may find on the eve of the exam that they were counting on you to read the book or take the notes, while you thought they were doing it. But the lack of pressure in a lecture class can be essential to the survival of underclassmen who have too much to cope with as it is and usually cannot handle a course that demands a lot of time and attention.

FRESHMEN PREMEDS

If you are a premed, you are determined. The first day of class, it will hit you just how many premeds there are, and that is just in your school—multiply it by the number of colleges in America. Do not worry—many of these people will drop out of the program as time goes by. The important thing is to try hard, study long hours, and not to be forced out yourself. This is hard, but presumably you knew it was not going to be a bed of roses.

There are a few practical dos and don'ts. Do NOT jump right in and declare yourself to be premed the first week of school. There is no percentage or advantage in this. Any course the registrar jives you about being open only to premeds is bound to be too competitive and advanced for you at this stage anyway.

Do NOT take your most important premed courses

during your freshman year, no matter how determined, capable and brilliant you are. You are not ready for the competition. Today, with the number of people applying for each place constantly growing, getting into medical school takes not only brains but really good grades. You are not going to get your best possible grades if you take important classes as an inexperienced freshman. If you do something this stupid, you will risk killing your chances for any sort of medical career, except nurse's aide or orderly or candy striper.

As a freshman and sophomore you should unofficially audit the more important courses such as organic chemistry and genetics. Take notes and go to all the classes. Do not just sit like a bump on a log. Pay attention. Most of these classes are big enough for you to even sit in and take the exam without anyone noticing. So take them—just do not sign your name. (You may even get your graded paper back when the professor throws it down on his desk in disgust saying, "Someone did not sign his paper." But pick it up when no one is watching.) You are taking these classes for practice. If you cannot get into the test, get a copy from someone the morning after. Do not think that you are the only one doing this. Look around you. Many people already have full notebooks the second week of class. They have been sitting in for the last two years and now, confident and pretty damned knowledgeable, they are taking it for a grade. Do you want to be graded on the same curve?

Premed classes are notoriously cutthroat. It is undergraduate competition at its worst. (A school of piranhas is preferable.) These students will not lend their notebooks, help each other out, or even smile. Do not bother calling anyone the night before the exam for an answer to a problem; they will give you the wrong one, if they bother to answer at all. You are not ready for this as a freshman or sophomore, as you can easily see. Unfortunately, most aspiring medics do not realize this until it is too late, and then it is all over. Realize it. Realize that premeds are more mature, more serious than the vast majority of freshmen, and they mean business.

REGISTRATION: HOW TO GET INTO CLOSED COURSES AND PROGRAM PLANNING

It often seems that registration is made as difficult as possible, especially for the freshmen, and especially if they want a particular professor or course. However, you should not despair, it is not all that bad. Believe it or not, registration used to be more difficult before the day of the computers.

There are tricks and methods to getting exactly what you want, and not what will keep the registrar beaming over balanced class sizes. But before we get into that, there are a few preliminary notes. As a freshman you must realize that you are on the bottom of the totem pole as far as granting class preference requests go. To put it clearly, the registrar feels that since you do not vitally (life or death) absolutely need a certain course to graduate, you should be happy with second, third, or even fourth best. You can take anything at the university (after all you have years left to go) and are just being obstinate. So if you give them a hard time, they are going to tell you to buzz off (in officialese, of course). If you really want to get into a certain course, *anticipate difficulty.*

The hardest courses to get into are seminars—especially seminars that majors need to graduate. If you have disregarded our previous warnings and still want to get in, it is fortunate that the system provides enough obstacles to stop you. Places in these petite classes are given to upperclass majors, then the underclass majors, then the upperclass nonmajors, and then you in that order. (And usually an upperclass major has to fight to get into his seminar.) To top it off, you must have the professor's signature on your schedule card, or whatever. This is nothing to someone living one subway stop from the school, but to the freshman living in Nebraska trying to get into a fall class at Berkeley it can be an insurmountable obstacle.

You can try sending the professor a postcard addressed to his office in the middle of the summer (but chances are he will not be there and even if he is he will not sign it). The best possible way to get into his class is to visit the professor the minute you arrive on campus, still carrying your bags, if need be. The professor, just like a prospective employer, will find it harder to say no to you in person than he would to a card, or a telephone call, or a face in the crowd on the first day of class. You have already showed such passion and enthusiasm for the subject that he cannot help but be swayed to agreeing.

But a freshman will not have the necessary prerequisites to qualify. We should not tell you this (it is like putting bullets in the gun for you), but what the hell—learn from your mistakes. You can get into a class without the stated prerequisites if you can convince the professor that you do have the necessary background. In the catalog it will list the prerequisites and then say "or the equivalent"; it is up to you to convince the instructor that you have "the equivalent." (This is also an easy out so the professor can decide who gets in and who does not and does not have to be bound by what courses they have had before.) Now if you are an upperclassman, you can offer much more concrete persuasion with the related courses you have taken. But in both cases, freshman or upperclassman, you should mention your extensive, independent reading on the subject—whether you have done any or not is inconsequential.

Aside from getting into the class before the term begins, you can always go to the first class meeting of the course, even if it is closed as far as the registrar is concerned. Many times people will drop right after the reading list is passed out, or as soon as they get a look at the professor. If you are on the ball, you can easily get a place in the class. The trick is not to give up too easily.

Finally, there are professors who will simply not accept more than X number of students in their classes. A major with only this last chance to take the course could lie down and die and he would still say, "I'm sorry but the class is closed." Your only course in this case is to graciously accept your defeat and to put your name on the top of the list for the next time it is offered. But here, if you are freshman, this is your opportunity to start some of your public relations work. Take another course with the same professor. Taking a course by a professor who was impressed but regretfully had to say no, is very smart. He is favorably disposed to you as an interested, informed, and intelligent student. He'll feel that anyone who is going to take several of his courses cannot be all bad. Professors are only human (although there are some incredible exceptions).

Today there are several schools who have become very enlightened about registration. They let everyone go to class—any class that they want—for the first two weeks of school, and then the students walk by the registrar and drop off their schedules. You can work out all your problems right in the field. And if there are overflows in classes, another section is created on the spot. It makes the whole process very painless and is a lot less trouble for everyone involved—or for the students, at least.

However, if you are not one of the lucky people at a school that operates in this fashion, and you do have to contend with the Big Computer two days before every term, take heart. We have words of wisdom for you too.

Today most schools use a "pre-registration" computer system. Preference sheets are passed out several weeks before a term begins, and before actual registration. You sign up for the courses and professors you would prefer. (Makes sense, doesn't it?) A few weeks later, you are told if you are in a course or "closed out." Of course there is the nightmare of being closed out of everything you requested. But if this happens to you, actually you are in luck! By looking distraught by this catastrophe, a sympathetic registrar or professor will generally "sign" you into classes (that is, enroll you in the class despite the class size limit). You may have to compromise and pick one or two other courses. But it's simply a matter of playing on the misery factor before the right audience.

If you are closed out of a course or two, you do have several alternatives. If you are willing to go to the trouble, you can get into the course. First, go to the first class meeting and see if the number of people who signed up for it have actually shown up. If they have not, get the professor to okay you, and you will get their place (a reason to make sure you go to your classes at least once). Second, during the registration day there is nothing that says you cannot sit in the computer room and run your course request through a million times. You could hit it lucky and get a place when someone comes in and drops it. This is called registration roulette. But the very best method is to go to the dean of students, the registrar, and the professor and explain to them how this is the only section of this course that you can possibly take because you have organized your schedule around your work hours. Nothing else will do. (This is a great method to use when you are trying to get into required courses in your major. However, if it is just something you want to take, play up the fact that you came to college to study this particular field, and now they are not letting you do that. But do not be belligerent, be distraught.)

Finally, the last word in computer registration is to know your computer (take it to lunch, buy it a beer). Know how it works. If they tell you that every time you send in a request sheet for one more course, you must list all your courses, do not think you will save time by just putting down the extra course. Put them all down, or you may find that the computer has wiped clear everything else and left you with only that one course. By the time you get a sheet filled out for all the rest, the computer may have given your places away. Too bad. A way to save time however, when trying to get one more course is not to send through sheet after sheet asking for just one more course, but to fill out the request sheet asking for several courses to be added. You will probably score on one or two of these, but if you happen to get into all of them . . . well, it is a gas watching all the buzzers go off in the room and to find out that you have thirty courses. Do not worry —you just get an armed escort over to the drop desk, where you drop everything you do not want and keep the ones you do.

One last word about registering. It is smart—if you can afford it—to take one extra course per term as a backup. Sign up for one more course than you actually plan to take. This way if you do get into trouble with a course or two in the middle of the term, you have the leeway to drop it without losing credits toward graduation. This is especially helpful if you are uncertain about a course. So if you are planning on carrying sixteen hours, take nineteen. Just before the drop date see what you are doing worst in and go ahead and drop it. You will not have lost anything. You will just be using a safety valve. If you had to drop a course you were counting on to carry you that much closer to graduation, then you will have to load up with an extra course the next time around to regain the lost ground (or worse, wind up on those five- and six-year plans where you go to college forever and ever).

The optional course should be in something that you are going to enjoy. It should provide you with a little relaxation, (the oasis in your desert of heated studying). To a premed bogged down with chem assignments, reading short fiction will seem like a sinful escape. It will not be drudgery unless you leave it to the last minute. But do not make your optional course some heavy reading class since you are going to fall hopelessly behind if you are really concentrating on another subject such as chemistry, biology, or physics. Who knows—you may not need to use it as a safety and you may finish it and *voilà* an extra credit toward graduation.

We cannot say it often enough: PLAN YOUR PROGRAM CAREFULLY. Aside from looking at yourself honestly and thinking about how much work you are actually prepared to do, providing yourself with safety valve courses, picking relaxation courses, etc., you should make sure that as you skip merrily along the road to D-Day, that you are picking up the necessary prerequisites that you are going to need to get into the more advanced courses in your field. For example, later as a sophomore you might want to take Course A, but it has a prerequisite of B and C or B and D; and Course B had the prerequisite of F. As you can see, the need for planning is evident. KNOW YOUR CATALOG and know your computer. This knowledge is more valuable to you than any of your textbooks when you are trying to run the gauntlet. Happy juggling.

2

DURING
THE TERM

Picking your courses and getting into them at registration is the hard part. Between registration day and finals, it is your ball game. You have time to take it easy. See a few movies, sleep late, drink, smoke, and indulge. And go to class every once in a while. Enjoy yourself during the term. You do not have to go to every class, you certainly do not have to read every book assigned, and you do not—or at least you should not—worry. Not yet. In most schools, even if you do not go to class once, if you score "As" on your exams and papers, you are going to get that "A." And we are going to show you how to do just that.

You must focus your efforts; that is all there is to it. To take an exam or write a paper and do it effectively, it is far better to know one thing backward and forward than to have a passing knowledge of everything covered. You can save yourself a great deal of time and endless trouble if you follow one rule: Do not attempt to memorize every fact presented to you or to read every book mentioned. Know only the most important information and know it well.

The chapters here will help you focus your efforts during the relaxing period after registration and before those horrid last weeks (and they are horrid to the grub and the cavalier alike). You should go to the first two weeks of classes. After the first week or so, you will know which classes require your warm body and which have seats reserved for the Invisible Man. You will also be able to feel out your new professors, and determine just how much effort each course will require. (The farther you get along in your career, the less time it will take you to make these judgments. Many seniors can accurately glean all this information in the first five minutes.) If you are planning on attending class regularly (my, but you *are* good, aren't you?), this is the time to make your excellent impression on the professor by asking interested questions.

If you do intend to go to all your classes, then by all means sit up front where the professor will be aware of your intense face. If you are not going to be dropping by so often, then try to remain anonymous. If you make an impression of any sort on the professor right away and then suddenly stop going, you will find out to your dismay that he is asking around about you. He may even send in a drop card for you. So sit in the back row and shut up. Do not get cold feet after the first few classes of something that looks as if you do not have a chance in hell of passing. Do not drop it. Hang in there and drop it *after* you fail the midterm or whatever. If you do pass the test, then you can always take is pass/fail. But we will go into that later.

It is really very simple. You must concentrate your efforts, decide which classes are worth spending your time going to, and which ones you have to go to, and then have a blast with the rest of your free time. The only thing you have to remember is that YOU MUST TAKE EVERY EXAM AND YOU MUST TURN IN ALL REQUIRED PAPERS. These are things that little grades are made from. So study for them and them only. If you do this, you will have a better time and probably better grades than any of the grinds.

Read on. The term can be bearable even in the classes you have to go to.

READING IN GENERAL: KEEPING AHEAD OF THE REST OF THE CLASS

On the first day of the class, the professor passes out the reading list (and an audible gasp rises in unison from the majority of the class while the rest faint dead away). It looks like the longest list you have ever seen. Half of the class rises and heads for the drop desk at the computer hall. Stay right where you are. No matter how long the list is, you are not going to have to read more than three books. Honestly. You do not even have to actually read these; you just have to know them. How can this be possible, you query, coming out of shock? We shall tell you.

First, lop off the books at the end of the list. The class will never get that far. (The professor just wants to have it all planned out, in case time stands still and he has an extra twelve months.) You might want to read the last one on the list, so that later you can mention it in an essay question. This has a very high "impress" value. It will appear as if you, avid scholar that you are, sped through the whole list. Rest assured— the last few books are just there for show.

Now that you have cut the list down to a reasonable size, go to your local bookstore. The books that are absolutely essential to the course will be listed there, and you will be expected to buy them. Here irate bookstore owners who get very upset about stocking books that no one ever gets around to using have forced the professor to delete a few more of the superfluous books from the list he passed out in class. List is getting shorter, isn't it?

From your list of essential books, cut out the ones you have already read. You do not have to read them again. If your roommate is halfway intelligent and has read some of the books, you can cross those off the list, too. You should be left with a grand total of about four books. You have sixteen weeks to read four books. Congratulations, You can join the Book-of-the-Month Club.

If you are one of those who cannot reconcile himself to these invincible truths, and you cannot see chancing an exam with such minimal knowledge, you can read several more of the books. In fact, you can read all of them if you eat sandwiches all term. But you are wasting your time. It is better to read one book and know it well than to skim all of them and get your facts wrong. If you attend class regularly and listen, you are going to be able to pick up whatever else you need to know for the essay questions on the books you have not read. The only place you might blow it would be on an objective test, but then you would probably get the questions wrong anyway even if you had skimmed all the books. You do not have to worry about objective tests in reading courses; the essays are the thing.

Plan your reading around answering questions on the themes and theories that are discussed in class. You do not have to read any of the books to do this. Listen to the professor. At some point during the term, he will mention several critics and their books dealing with the authors, and the themes you are studying. Pick these books up at the library. You do not have to read the whole book. Read the first few chapters that deal with the critic's point of view. Then pinpoint your assigned author and book in the index. Read everything this critic has to say about your author and his masterpiece. There you are. If your professor disagrees with this critic, you can write a long diatribe against the poor man. If your professor idolizes the critic, then your essay can dwell on the man's astounding perception and astuteness. (This is called indirect praise—use it.) Here you have an instant essay on paper which sounds very learned—with almost no reading involved.

See how easy it is? Just cut down the list to the bone and then concentrate your efforts. Reading is one of the few things in college that can grind you down and then plow you under. One Princeton student we know did a marathon reading of *Anna Karenina* in ten hours at one point in his career, and he will never forget it or dear Anna. This sort of thing is needless. If you read carefully, plan carefully, think about the little of the reading list you do get around to, and then supplement your knowledge with a critic's theories and your class notes, you are going to be in great shape. You should have no trouble writing essays or papers or keeping up with the class.

MONARCH NOTES, CLIFF'S NOTES, CLASSIC COMICS, AND MOVIES

Despite the preceding chapter, you are nervous about the reading list. You just want to feel better about the five books you did not read. Or you just did not bother to read anything. It is the eve of the exam and you need a quick way out. Okay. Just as there is a form of instant paper, instant masterpiece, and instant thesis, there is a form of instant knowledge. You have several options still open to you.

In times of trial, most people turn to the notes—Monarch or Cliff's. They are not bad. They could be better. To make it crystally clear, they both have their flaws and they are glaring. If you have a choice, try to stick with Cliff's Notes. Monarch Notes—especially the Shakespearean series—sometimes espouse theories that were junked in the early forties. Themes are not explained as clearly or as concisely as in Cliff's Notes. Both types contain sample essay questions in the back. Forget them. These are never asked beyond the junior high school level. Finally, the plot summaries and interpretations were designed to help the younger reader understand. Many times the interpretations given for chapters and scenes are going to be too naive for you to use without sounding like a dummy, or without showing that you obviously did not read the book.

Aside from these little problems, the Monarch Notes and Cliff's Notes both contain excellent plot summaries, and in most cases, they cover the essential subplots as well. You can count on the notes to pull you out of a ditch as far as objective knowledge of the author and his novels go. You are putting your life on the line if you use the notes' subjective interpretations for your essays.

Many people swear by the Classic Comic series. Do not laugh. There was a senior at a well-known Catholic college situated in Washington, D.C., (who wishes, understandably, to remain anonymous), who owned the entire series. He swears that he never would have made it through his English major without them. Classic Comics are good. They are easy on the eyes, short and pithy, contain the plots and essential subplots, and you can get a good idea of the interpretation of the scene through the expressions on the character's faces, naive or otherwise. As far as factual information goes, Classic Comics are in the same league as the notes. They are more interesting than the notes any day. The only reason they are not more popular is that they are comic books. Who is going to admit that they studied from comic books? You do not have to tell anyone, however. Use them. They are cheap. They can be hidden easily.

Some people rely on movies. Do not be a fool. Plots are changed, happy endings are added, superfluous characters die, and stars have entire roles rewritten for their personas. One Brown student—a sad, foolish boy—came home to visit his parents and to watch *Anna Karenina* (this book really gets around—everyone has to read it at one time or another) on public television. He went back to school confidently.

Later he wrote a glowing essay about Anna, and how she went on to marry Vronsky and live happily ever afterward with him and her son. Only later did he find out that Greta Garbo was supposed to have thrown herself under a train. (The movie was originally released with the proper ending, but it was too sad. Since the public wanted fairy tales, the producer tacked on the new ending.) Do NOT rely on a movie. NEVER, EVER.

There is a surefire method, safer and more sophisticated than your notes or Classic Comics, and far more reliable than a movie. Go to your library and find the *Masterplots*. *Masterplots* is a series of books—an encyclopedia, actually—containing the plots of all the notable literary efforts of the centuries. It covers just about every book you will ever have on a reading list (fiction that is). Ask the librarian where to find this treasury of knowledge. If your college library does not have it (some do not for the obvious reason that it is a sophisticated version of the notes), try the local public library. Pinpoint your book and read the summary. The *Masterplots* will have a short outline of the plot, with all the subplots—essential or not, all in one or two pages of very tiny print. You will have to come up with your own interpretation or get one from another source. (You could get a book of criticism on the author in question. Find the chapter dealing with the book you supposedly read. This will outline the themes of the book and focus on the most major of them for you.) For a little extra power, you could read the editor's introduction to the copy of the novel you have. Go over your class notes, and you are all set. You are better informed and able to write an essay that will seem more informed than the poor slob who suffered through the whole book.

CLASS ATTENDANCE: THE WARM BODY METHOD

College is not like high school: whether you go to class or not is entirely up to you. Many people do not go often or at all. So be it. In large lectures it does not matter whether you go or not. No one is going to notice whether you came, stayed home and played cards, watched television, or got stoned. Some professors try to discourage this by passing around a sheet of paper that everyone is supposed to sign. If you do not think you are going to make it to many of the lectures, go to the first one or two classes and find a diligent type who will be willing to sign your name. Sometimes you will have an obliging friend who will do it. (For the worriers, don't sweat it—no one checks handwriting.)

On the other hand, there is a lot to be said for attending class regularly. Aside from the fact that class lectures and notes will relieve you from doing much of the reading, you will make your cheerful, dependable little face known to the professor. Here we get into the Warm Body Method. The Warm Body Method is very simple: you fill a seat with your warm little body every class meeting. You do not ever have to contribute to the class discussion. Just be there. You smile at the professor. Look interested. Try—it will come with practice. Let him know that you are there. If you are called upon for an answer or an opinion, while practicing the Warm Body Method, you can get around it easily. Just say, "Well, sir, I wasn't exactly sure of that point myself," or "I was a bit confused by that passage, or problem," etc. When grading time comes around, the professor will be predisposed in your favor. After all, you came to every class; obviously you were working. You were definitely more interested than those who never bothered to come at all. By the end of the term the professor will know you, even if you are convinced that he does not know you from a hole in the wall. The originator of WBM filled a seat in Russian Literature for a semester and was pleasantly surprised when the professor greeted him by name at a newsstand. (He vouched for WBM recognition only in the class settings)

Attendance in a seminar is another ball game entirely. You almost have to go. *Almost,* we said. If you do not go, many professors view it as a personal insult. These seminar classes are very small. Your lack of scholarship and dedication is going to be noted. You are taking your chances if you do not show. So go. Fill the seat. You will find that most of the rest of the class is WBMing it too.

Going to class regularly has its advantages. (Staying home in bed, alone or with others ain't bad either.) But by going, you will know what is going to be on the tests, what material is being covered exhaustively, and what is being skimmed. At the best you will not have to read one book—just review your notes for the exams. At the least, you will meet some interesting people to study with.

SLEEPING IN CLASS, CIGARETTE SMOKING, BEER DRINKING, AND OTHER INTERESTING WAYS TO PASS THE TIME

Obviously, every class you go to is not going to be the most fascinating and stimulating hour you have ever spent in your life. Some can be ghastly clock-watching sessions where your whole body falls asleep and your mind numbs. If you are going to be a firm follower of the WBM, then you are going to have to devise a way to entertain yourself quietly and unobstrusively. There are ways to do this. Look around. See those people over in the far corner? They certainly don't look bored.

You can sleep in class, and it is a good way to WBM painlessly. But you have to watch it. Flopping your head down on the desk and suddenly snoring moose calls is going to insult the professor. (Professors tend to have very sensitive ears.) This kind of blatant disregard for his eloquence is going to insult him more than if you called him a windbag to his face. There is a definite technique involved in sleeping away a class hour. Sit on the side of the classroom, or in the back. Do not put your head down; it is a dead giveaway. Tilt your head away from the professor and snooze softly away. It admittedly takes practice (unless you are a born bird and can sleep with your head tilted at any angle). People who wear glasses have a definite advantage because they can tilt their heads and let the light flash off the glasses. No one can then see that their eyes are closed. It does take practice, but you have four years to perfect your technique.

It takes about four and a half minutes to smoke a cigarette. Half a pack can kill the hour, and you can practice your smoke rings if you sit unobtrusively in the back.

Beer drinking can really liven up the back rows of cinema or art history classes when the lights are low. Other rear-row diversions are readily available in these classes.

You can do puzzles and play paper games by yourself (or with a neighbor—an hour a week for sixteen weeks can mean a lot of tic-tac-toe and it can get very boring). You should try to do something creative that can later be used as a wall decoration back at the dorm. One USC coed worked for two years on a mural done

on pieces of typing paper. It was begun appropriately
enough in a cinema class. She started with:

```
                K
                ALANLADD
                T           O
                H           R
                E           I
        GINGERROGERS
                I           D
        DAVIDNIVEN          A
A               E    O      Y
L               H    E
A               ELSALANCHESTER
N               P    C
B               B    O
A               U    W
T               R    A
ERROLFLYNN          RONALDREAGAN
S                   D
```

This was the final product of her first hour. She worked
it out from there. At the latest estimate, the mural
measured five by eleven feet, and contained every actor
listed in *The New York Times Directory of Film* (and
none were used twice—the mind boggles). Now that
is creative. You can play this game with politicians,
authors, scientists, or whatever your class and subject calls for.

There are some more obtrusive ways to entertain
yourself in class. You can ask a "profound" question.
This works the best in a seminar. In every seminar,
aside from the professor, there is a loudmouth who just
never shuts up. Disagree with this idiot the second he
opens his mouth. Do not come right out and call the
fool an asshole; just disagree with his point and nail

the bastard to the wall. It is an unwritten rule that once attacked, this loudmouth will go on to greater stupidities, and the quiet class that has had to listen to this clown for the whole term (now tasting his blood) will rise to the occasion (and finish the kill). A splendid argument will ensue. The professor will be pleased that someone else is talking for a change, and you will win brownie points for a contribution. It is kind of fun, too. Once you get the action started you can sit back and enjoy the ride.

Seminars have traditionally been the best atmosphere in which to write your weekly letter to Mom and Dad. If you did not learn basic in-class letter writing in high school, we will review the essentials for you. Keep it simple. Do not head your paper, "Dear Mom" and begin to write under that. Write the body of the letter in class, and look as though you are furiously taking superb notes on every golden word spouted by your illustrious professor. (A few awestruck, adoring glances in his direction are quite appropriate.) After class, add the heading and the closing, address the envelope, and lick the stamp. Never do this in class; it is simply not good form.

Now, if none of these diversions is entertaining enough, you can always leave the class. It is not like high school, where you have to raise your hand and get permission; that would be like needlessly pulling the emergency brake in a speeding train. Simply get up and leave. But do not yawn, show you are terribly bored by all this nonsense, and stalk out. Oh, no! You must leave in style.

One of the old favorites—and still one of the most effective—is to look sick. Do not overplay it and put your hand over your mouth and rush out, books flying. This kind of acting went out when they developed sound. Just wipe your forehead, rub your eyes, and look fixedly at the closed or open window as if you were trying to get some air. After a few minutes, start taking short, shallow gasps—unless hyperventilating makes you faint, gather up your books, and leave.

This has an offensive rating of zero, and a sympathy rating of ten. But you can do it only once or twice. (If it becomes a weekly drama, the professor might suggest that you see a specialist).

Or you can suddenly look at your watch and quietly rush out. The offensive rating on this, however is four. It is chancy, but you can always tell the professor later that you had a doctor's appointment (very convincing especially if you pulled the "oh, dear, I am going to be sick" routine the week before), or a meeting with the dean of students.

Finally, if you know you are not going to last the hour, or if you really do have to leave for some legitimate reason, do not sit in the middle of the class. Sit near the side. If you can possibly get the first row, first desk, you will be able to fold your tent and steal away without any act, in a large class. (If it is a large lecture class, it would be better to miss it entirely than to risk the professor's hurt pride by leaving in the middle.) In seminars, you can usually disappear at the break, never to be seen again.

PROFESSOR RELATIONS AND GENERAL PR

Professors—most of them—are human beings, just like you and me. They like to be treated as such. This is what you have to remember during the term. They do not like to be insulted, and yet they do not like to be fawned over. They like nice, quiet, INTERESTED, considerate people. For the most part, they like the student body—(it is just that they like some bodies

better than others). So there are certain things you
should do, certain things you should not do, and certain rules of etiquette you should follow.

1) Be considerate of the professor. No one likes to
spend his weekend reading papers only to have one
clown show up with his two weeks later. You are
going to get a lower grade on irritation value alone.
If you need more time, ask for an extension for the
project, paper or whatever. Then get it in on the
new date. Do not take advantage of the situation or
the professor.

2) Do not be the class asshole and monopolize class
time with your opinions. Professors like contributions,
but do not like little pseudo-professors lecturing from
the back row.

3) Do _not_ be the person who constantly turns the
seminar over to small inconsequential points, while
the professor tries vainly to get back to the matter
at hand. If you do, your arrival will be greeted with
a small groan at every class. (That gleam in the professor's eye when you open your mouth is not joyous
anticipation but rather a suppressed desire to feel his
fingers around your throat.)

4) When you are trying to bluff your way out of
something (such as answering an unexpected question), make your bluff short, and always end it with a
question.

5) Try to get to class on time. Professors do not
like people waltzing in late and leaving early on a
regular basis. It upsets their rhythm. If you do show
late and disrupt your seminar or small class, then
always apologize after class.

6) Class clowns are really out of place in college.

7) Office hours are for extra help and small inconsequential but interesting points. It is smart to see

each of your professors at least once during the term, to help him place the face. If it is a large class, then he will remember your name, and he may remember that you were interested enough to drop by and ask a few questions. This will be invaluable when it comes time for him to give you a grade—especially if your grade is teetering on a borderline.

8) But do not turn into one of those who is constantly dropping by to spend a few hours chewing the fat. (He will feel as if you are trying to drain him dry.) Most professors do not like keeping office hours precisely because of these types. Take the hint —don't become a must to avoid.

9) No matter what happens, you are the one who should be pleasant and slightly apologetic. If you are in a small school, finding yourself bad-mouthed to the faculty is not a pleasant experience.
10) Always consider an invitation to a party, a coffee-shop, or a faculty tea as a command appearance. (The queen could have you beheaded literally; your professor can do so quite as effectively, figuratively.)

11) Finally, becoming a particular professor's groupie has its pitfalls. While you may find the professor one of the most fascinating and sexy people you have ever met, BEWARE. Most professors have groupies. Many have joined the total-honesty encounter movement. Rather than getting a high grade from your close friend, you may find yourself getting the low grade you deserve. After all, honesty is the byword in all of today's relationships. So save your admiration for the term after you have taken the professor's course and gotten your grade with the rest of the faces in the crowd. Then, you can become buddies, and chances are he will write you a decent recommendation, remembering only that he gave you an "A" but not what for.

3

EXAMS
AND PROJECTS

Exams and projects conjure up memories and/or visions of all-night marathons of reading and writing, and building little Medieval French villages. Don't kid yourself—this will happen to you. There are very few people who manage to make it through school without having pulled at least one night of agony somewhere along the line, and these people should shine as paragons of scholarship to all the rest of us who cannot do anything until the pressure from the last minute builds up to an explosion of frenzied panic. Sadly enough, you can stay up all night, burn out your eyes and brain, and still get the worst grade in the class. The reason for this is very simple. Most people when faced with the do-or-die situation in the morning, will stay up all night and attempt to cover a whole term's work in ten hours.

If you stagger into an exam hall at nine o'clock in the morning after reading four novels, three articles, your textbook, and your notes, you are not going to be ready for the exam. You will be a basket case. This is not the way to prepare. The key to last-minute preparation is concentration. It is careful evaluation of the type of knowledge that your test will require and the systematic study of that—and only that—knowledge. You cannot develop your own strategy to overnight success overnight. It takes two or three years of actual field experience on the academic battlefield. We have gathered a few hints to help you in organizing yourself for that last moment of panic the night before, so that you will not be sorry the morning after.*

*There was total disagreement between the authors on pulling an all nighter the night before an exam. (It is agreed that an all nighter is feasible for papers and projects.) One advocates utilizing every minute before the exam to study and going straight from textbooks to bluebooks. The other is vehemently against going in to take an exam without having had any sleep for 24 hours or more. The authors recommend that you

EXAMS

No matter how carefully you plan your courses and try to get into paper and take-home exam courses, you are eventually going to take a course where you will have to take a sit-down exam during a time limit, in a classroom. There is no way around it. And there is nothing to fear. If you know how to study for the type of exam you will be given, and how to take it, then you have nothing to worry about as far as a grade is concerned. The amount of material you know and have at your fingertips is worth about 20 percent of the grade; the rest is determined by your studying and test-taking expertise.

The Multiple-Choice Objective Exam:

Studying for this type of exam is the easiest. Basically all you have to do is connect the important words, phrases, and notables in your mind. You may know nothing about geology, but when you take the exam and see "many-faceted" and then have enough presence of mind to connect with "diamond" on the line below, you are in. You should study the words and phrases, and see them only in connection with each other. This method will also help with matching columns—which can be a bitch. If you miss one, chances are good that you will get many of the others wrong. So concentrate on connecting and not on memorizing. All the answers

---

cram all day and evening. Then go to bed and get a good night's sleep. For many, sleep not only means a refreshed body but seems to give the mind time to sort and file much of the precious crammed knowledge. One other note of interest: many people find that if they eat just before going in to take an exam it leaves them sluggish. Find out what works best for you.

are going to be right in front of you; all you have to do is zero in on the right one.

After you have the basics down, trot through the text and your notes to learn the basic theories, so that you will not mess up on the true and false. Then zip through one last time to pick up any unique and interesting facts that are really just points of interest (in other words, footnotes of history). Professors love to put these on tests because they will be able to see if you "actually read" the text. But don't worry; most of these unique little facts are the things that you will remember long after you have forgotten the rest. (For example, many learned professors and medical experts theorize that John Milton might have gone blind from syphilis? True. Now, who is going to forget that?)

Identification Exams:

Identification exams are also very easy if you know what you are doing. There is no way you should not get every point for the ones you can identify, and at least partial credit for the ones you do not know from Adam. The method is simple. First you identify the person or thing by name, then date it. If you cannot come up with a specific date, approximations do count. Then tell why it is important, where it, he or she can be found and you are finished. For example: Identify Norman Mailer. (5 points)

Norman Mailer is an *American writer* who came into prominence *after the Second World War*. His *first book* was *The Naked and the Dead*. Recently Mailer has concentrated on *nonfiction* and has written books on *Gary Gilmore* and *Marilyn Monroe*. He has also run for *public office*, made several films, and dabbled in *boxing*.

This identification was worth five points. You put in five facts. One for every point. The man is identified, dated, what he had done to be noteworthy is mentioned, and what he has done lately. You get all the points.

Now suppose you cannot possibly come up with the required number of hard facts. Do not panic. If you know anything at all, write it down. For example; Identify Bismarck. (5 points)

> Bismarck was a *German general*, who lived in the *nineteenth century*. He was often called *Blood and Iron*. His first name, however, was Otto. [This is called stretching your knowledge.]

So now you have used all your actual knowledge of Bismarck and it is not enough. Add two more "facts." His leadership had a profound effect on the face of the European continent. (For all you know, he might have or he might not have; but chances are that if he is on your European history exam, then he had some effect —profound is an immeasurable and it sounds good. If you are wrong, the professor will merely think you were mixed up or misinformed rather than totally ignorant.) His wife was an Austrian. (Who knows? Who cares? And the professor probably will not know either if you make this tag fact obscure enough. He will wonder what you know that he doesn't. Of course it will not do you any good if he is a delver and spends time to check it out, but you lose nothing either way.)

There is only one drawback to an identification exam. Some professors get cute and put in obscure people or things that there is no way you would know from the course. This is usually found only among very young professors or on exams that graduate students make up. Usually these are not counted against you; they just make life frustrating. For example: Identify Frank McHugh. This was on a history exam in an Ivy League school. No one knew who the hell he was but most made brave responses. The teacher (who was a grad student) thought it a great joke, getting his kicks out of throwing many students off balance. Frank McHugh was a character actor for Warner Brothers. He appeared in many films of the thirties and forties. Most notably he often portrayed James Cagney's sidekick in such movies as *The Roaring Twenties* and *The Crowd Roars*. Big deal—so who cares?

Short-Answer Exams:

You probably will not have to worry about these "fill in the blanks." They are rarely given these days because of the furor that arises when they are graded. But in case you are faced with one:

This type of exam is a bit harder, but once again, remembering the key words and phrases is the way to an "A." However, here you will have to remember how to spell the key words and phrases, and that entails actual memorization. When you study for this sort of thing, you should not spend long hours studying the complicated passages and theories. The questions will be pithy, so center your study on the pith of the matter.

Essay Exams:

"It is important to write a lot. Most professors will tell you that they want quality of thought rather than quantity of blue book. Don't kid yourself."

—J.C. Lehigh, '75.

The essay exam is the one type that you are going to meet over and over again. It begs for bullshit. By the time you are a senior, you will have it down pat and recognize the basic truth of the above quote. Most of the questions that you are going to be asked to answer in the essay form could be answered in a sentence or two, but they want an essay. Give it to them. Write, write, write. Write until the hand begins to cramp and shake. Then write some more.

What you write is important, but to take an essay exam you do not have to have actually read everything you mention in the essay. Just mention the books you have not read in a vague way, and concentrate on the ones that you did read. This can be done easily by writing, "of course one can see this theme in so-and-so's (author) BLAH-BLAH (title) and in what's-his-name's (author) SUCH-AND-SUCH (title), but for the sake of brevity, it is most clearly illustrated in THE ONE BOOK I DID READ by THE AUTHOR." Of course if you did read a second book by all means say, "The same is also illustrated in the works of AUTHOR #2 in his WHATEVER." From what you did read, draw parallels and mention characters; if you can paraphrase a line or two from the book, it is helpful and all to a good cause. If you know the names of characters and the plots of the books you did not read, you can always throw those names and events around, too.

Now as far as writing about the theme of any book, one should forget fact and concentrate on TRUTH. In essence, one must deal with general profundities, and universal TRUTHS. For the sake of brevity, we will list them.

THE TRUTHS

Man struggles against: MAN	(history, poli.sci.)
FATE	(history, literature)

NATURE	(geology, literature)
HIMSELF	(psychology, literature)
TECHNOLOGY	(urban studies, lit)

And there is always LOVE: of anything, anybody, and everybody.

Finally, the most important thing to remember is that few professors really read your essays. What they will do is scan it, and what they scan for are the "buzz words." What are the "buzz words," you may well ask. Buzz words are the theories, books, and authors mentioned during the lectures. You do not really have to know anything about the buzz words in your particular subject; just pepper your essay with them. This will show your instructor that no matter how long and detailed your essay is, it merely touched the tip of the iceberg of your knowledge.

There are some poor souls who cannot write essays to save their lives. They seem to have been born with this handicap. A sad story points up this fact of life. This is the tragic tale of a coed at Marquette in Wisconsin. This poor dear, aware of the difficulty she had in English, managed to buy a copy of her English final. It was an essay exam. She pondered over the questions, read the pertinent books, and even went to the library and took out relevant critical books on the assigned books. She wrote several sample essays for each question and tried to memorize the ones she thought best. When she finally went to the exam it was with a confident heart. She got a "D"—from a professor who was known for not failing students.

This story points out only that it is not really how

much you know or how little, but how well you can express yourself that counts on an essay exam.

There you have it: a flowing pen, full sentences and good grammar, basic plot summaries, buzz words, and a bag full of profundities—in other words, bullshit.

The Take-Home Exam:

While it sounds as if you cannot miss, you can blow this one easier than any of the rest. Remember; the professor gave you this exam early so that he would get coherently written essays rather than hazy, panic-stricken thinking, so your exam has to be clear, concise, and littered with quotes. (It also has to be long.) Any bullshit you sling on a take-home has to be artful, exquisite, and smooth to go over. Remember; nobody is actually going to spend the three hours that the professor said they should on the exam. (Although it may take some only three hours to copy their essays into the blue books.) Once the grinds go home with that exam in their hot little hands, they are going to spend every waking moment working on it. So you must spend time to do it.

Although the popularity of take-home exams is calming down after its surge of prominence in the sixties (mainly because professors were getting full-length papers instead of concise essays), be aware that it is still used often, and keep up with what's happening in classes you do not attend. At least keep in touch with someone who does go to the class regularly or attend the last two sessions before exam time to be certain that you do not miss out. The sinking of the *Titanic* is minor compared to the sinking you are going to feel when you walk up to the teacher to ask for your copy of the exam and he looks at you in surprise and informs you that he passed it out two weeks ago. Death is easier than the realization that while you were boning for comprehensive surprise essays, everyone else had the exam at home. Be there when they pass the exam out, or know someone who was.

The Oral Exam and the Art of Conversation:

Taking an oral exam can be a harrowing experience, and if it is at all possible, try to avoid them. (Better you should get the plague than have to take an oral.) The professor (or Inquisitor) will ask you a question and you must answer it. (You will feel paranoia—but as they say, it is not paranoia; they really are after you.) It is tough because it is harder to gloss over much, when they can stop you at the first sign of vagueness, but if you are skilled in the art of conversation, you have a chance. When the professor asks the question, think it over. Do not start right in. Then talk about what you do know . . . and talk. Do not let anything get you off into uncharted waters. Pretend that you were absolutely fascinated by one certain book (the one you did read). If the professor tries to get you going on something you did not get around to reading, say "Well you know when I read that . . . I really was not sure . . . and I hate to confess this, but I did not understand that author's point (or that theory or interpretation or intention—whichever is most suitable.) " If you are asked about specific plot sequences in the unread book, say, "Yes, but can I ask you a question?" Then ask a question that leads back to your territory, or about the connection of the two writers and their books. Why were the authors of the time so perplexed by this moral problem, was it something pervasive in the moral atmosphere of the time? Do not let them lead you onto thin ice. Talk of nothing unless you are sure you know what you are talking about.

Talk about anything except the fact that you have not read the material or that you read it but did not understand it—unless you have to use this second fact as a ploy to turn the conversation back your way. This type of exam can become a gossip exam if you are an expert, but if you are just good, you can turn it into a discussion of the relative merits of the course as a whole.

If you bomb, you bomb. We warned you.

HOW TO TAKE AN EXAM ON A BOOK YOU HAVE NEVER READ

You read three things on the list. The professor asks you to write an essay question about the one you did not get around to. Allow yourself one full minute of despair. (Just enough to get the old adrenaline really surging.) Then starte to write, write, write, write.

Most nonscientific exams are essay types. This should present no problem, since essays are bullshit tests. You just have to spread the bull around a little more carefully. It is very simple.

Start your essay with a statement taken from the question. This gives you a half a page in your first blue book and assures the professor (if somewhat falsely) that you are going to stick to the question asked.

Example: Why were the Middle Ages considered moribund? You begin: The Middle Ages have always been considered a moribund time. Essentially this is a correct assumption. Stretch this out a bit and you have already managed to fill up the first page of the blue book. See how easy it is.

Now to get down to the basic facts, which you do not know. Since you cannot give quality, you are going to have to give quantity. (Don't worry—many professors cannot tell the difference anyway.) If you have been to class, read the books, read your notes, if you know anything at all, use it. Blow it up. Take it to the limit. Rack your brain—it is undoubtedly a fund of odds-and-ends ideas that you can work your way to, and just keep writing. Branch out from that little bit that you do know to something you know a lot about. Quotes from classic literature (i.e. Shakespeare, the Latin writers, etc.) are always helpful.

Now that you have gotten the general idea, here is a question done step by step.

QUESTION: What is the function of technology in *1984?*

You lead-off: "Technology plays a very definite and important role in George Orwell's *1984.*"

(Obviously this is true or the question would not have been asked—or more accurately the professor thinks it is true, and on bullshit answers you are better off agreeing.)

Your general profundity: "It illustrates the classic dilemma of man versus machine . . ."

(This can fill up innumerable pages as you diversify and expound on this dilemma.)

From class notes: "The two problems they tried to solve were how to kill vast numbers of people with no advance notice, and how to find out what a person is thinking against his will . . ." which leads you to

General profundity #2: Man versus man. (A master bullshitter could write volumes of criticism knowing only this much. So you see you have no problem—you need only fill a couple of blue books.)

Your buzz words: tyranny
                  totality
                  totalitarianism
                  liberation
                  freedom
                  power

Your "classic" quote: "What a piece of work is a man, how noble in reason, how infinite in faculties; in form and moving how express and admirable, in action how like an angel, in apprehension how like a god; the beauty of the world, the paragon of animals." (HAMLET, II, ii, 300–04.)

Your figure from mythology: Icarus. "Like the fabled Icarus, the hero of *1984* stood out, was an individual, and was destroyed at the height of his symbolic flight of individualistic thought and action."
(You need bend almost any mythological figure to suit your needs. Just think of one and give it a new interpretation. You will get credited for creative thinking.)

Obviously to get all this in you have wandered far afield through hither and yon. But the TRUTHS, the quotes, the overall grasp of the staggering concepts of the book that you convey will show the professor that you have not merely read the book thoroughly and are attempting to feed it back to him, but that you have pondered long and hard about it. His course—and particularly this book—have obviously had a deep effect on you and your life.

So write, and write your heart out, as if your life depended on it. Get yourself into that desperate drive, and write, write it all as if you know everything, too much in fact, you just cannot get to the point, and oh, dear, hurry, you have just run out of time, and there is so much more to say. . . .

CHEATING

Despite honor codes (which actually may be responsible for some cheating by making it so temptingly easy) and moral feelings, cheating goes on, and on, and on, and on. At one time or another everyone is at least tempted. (Then there are those few who have cheating on the brain and spend all their time devising devious methods of cheating on their library slips.) So do not be horrified if the thought should appear unbidden in your mind, or in that of a friend. But be practical about the whole thing: cheating is impossible on essay exams, impractical on detailed short answer, and very dangerous on objective tests. (That leaves take-homes, and even cheating freaks are stumped on how to cheat on a take-home.) Remember: if you are caught cheating, you are done for. Therefore, cheating should be resorted to only in the final depths of despair. If you have sunk that low and there is really no other way out, here are a few of the better suggestions and hints to guide your cheating.

Peeking at other papers around you during an objective exam is not only stupid but unreliable. (So who says Joe Magilla next to you is any closer to the right answer than you)? If you are going to cheat, and risk getting caught and destroying your college career, than at least have some class, a little flair and sophistication. Remember; not only does it gives you a stiff neck, but peeking and craning went out in the fifth grade.

One of the better tried-and-true methods which is still popular is to have an "A" caliber classmate drop his worksheet complete with the correct answers on the floor near you. This can be risky. Someone else's foot could get to it before yours does. This is not the major difficulty though, since you can always practice to perfect your lightning foot; the biggest obstacle is finding an "A" student who will part with his precious answers, since most of these students are so jealous they make Othello look tame.

Another alternative is to pay someone—preferably someone who is smarter than you are in the exam subject—to take said exam for you. This works only in very large lecture classes, and even then there is the element of risk. Remember: if you went to class even occasionally, the professor saw you a few times during the term. There is always the chance that he will collect the papers himself (professors get very sentimental around exam time) and when he collects the paper with your name on it from a strange face, the steel-trap mind will snap and suspicion will follow. For your information, this is where Ted Kennedy blew it at Harvard. (We said everyone. . . .)

Coughing codes (what do you do in the winter when half the class has colds), passing answers (which are likely to be hijacked), and relying on others in general is very risky. Obviously, these are not the best methods of cheating. To err is human, and nine times out of ten your fellow criminals will err. And you will not feel very divine after screwing up the exam. If you are going to cheat, it is really best to cheat all by yourself. It is clean, reliable, and just less messy and chaotic. There are two basic plans of attack you can take. They are the bathroom plant and the cribsheet routine.

Bathroom plants work very well, if you can remember the facts while you race back to your blue books. They are a little more sophisticated than most methods, and very few professors or proctors expect anything like cheating that is well thought out and planned. Its best feature is that it is very difficult to be caught red-handed. In actual practice, it is very easy to pull off and can even be used for notes on an essay exam. Write out vital notes and put them in the john that you will use during the exam (preferably in a plastic bag for protection). Do not leave your notes on something that might be ripped off while you are working away in the exam hall. An unattended sweater is likely to disappear. (College students are not supposed to cheat, but no one denies that they may steal.) Write the notes on a small piece or pieces of paper (depending on how little faith you have in yourself) and tape them behind the toilet tank or whatever *in the stall* (shades of James Bond). Most of the proctors will not follow you into the bathroom, but you never know. They might be perverts.) You certainly would not want them to peek in to see what is taking you so long, only to find you in the midst of pulling your notes off the bottom of the sink. (Cheating is a serious business, not a Marx Brothers film.)

If the proctor is the responsible type and at all reluctant to let you go, you can still get out. Women have it easy, they need only to whip out tampons (or if you want a greater response, a sanitary napkin) and look very distressed. No proctor in the world would refuse such a request. (Your prop can also be a terrific stash for your plants.) Response is not as readily granted to guys, but they can pull the stiff-upper-lip routine by saying that they ate a huge breakfast and now the old nervous stomach is acting up. If done convincingly, this will get you out of the classroom so fast that your head will spin. Remember: you must come back from the john looking vastly relieved, but this should not be difficult since your problems will all be solved by the trip if your plants are good.

Crib notes are great. They are probably the most widely used form of cheating. (It is theorized by some that George Washington was a crib user.) Here the act of cheating is refined to an art. Imagination, innovation, and a talent for miniaturization are the bywords among the crib note fans. The most experienced cribber can distill the knowledge of a term into three or four little papers and roll them to fit inside a clear Bic pen. A flashy pro from Brooklyn College reports that he *indexed* his crib sheets and fanfolded and taped them to his ruler. He explained that this was done so neatly that when his ruler was laid on the exam table, not even the person next to him was aware of his little encyclopedia. (Someone should start a crib-note museum.) Experts can put the New Testament on the head of a pin and make it legible.

However, for the novices among us, here are a few suggestions. Crib notes can be put on or inside almost anything: watchbands, pens, bracelets, big lockets. Just write tiny crib sheets and hide them in rolls of Kleenex, or packs of cigarettes, in matchbooks, cough-drop boxes or gum packages. You can also tape (or if you are Edith Head you can sew or embroider) notes to the inside of a skirt hem or a sweater, which you will take off halfway through the exam.

Finally—and not least of all—there is the ever-popular body method, since you can be certain you will have your body with you. You can write on the inside of your arms under long sleeves. (The back of your hand is out since the professor or proctor might think something was strange about taking an exam wearing gloves). However, if you have an unquenchable thirst to join the circus as the tattooed man/woman, you can skip the traditional crib method of basic formulas and shorthand. Go all the way. Cover your body with everything you know. You will have to go to the bathroom to read most of them however. Just make sure that the most necessary are the most accessible.

Remember though, cheating is very impractical, dangerous and it really does kill a good deal of the sport and challenge. (After all, aren't you going to college for the sake of the game? Does it matter whether you win or lose? If you believed that, you would not be reading this.)

THE CURVE

Some people are born test takers and never fail to do well. Others, remembering they got a "B" on the math test in the third grade because the teacher graded on a curve (and Mary the Brain was out sick that day), have come to rely on the proverbial curve. You are a fool if you rely on the curve. In college, the grinds are never sick on exam day, and the median grade is never a "B." There is always one clinker in the crowd who will pull down the only perfect score in his life (may he die by the pen) and leave you at the bottom of the heap.

For those of you who have always wondered just what the hell a curve is, (it is not the outline of a female figure that is being alluded to). Read on:

The most common curve used is the bell curve against which the grades are graphed. (Of course, there are those graph lovers who will plot a set of grades on several different curves just for the fun of it.) But on your basic bell curve, with the lowest grade a 15 and the highest a 95, the picture would be like this:

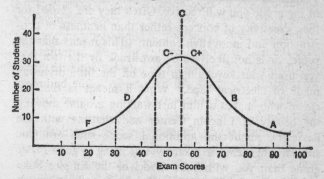

Average score on the test: 55 points = C

NIFTY DRAWING OF A BELL CURVE

The bottom ends of the curve are at 15 and 95. The 15 and a short way up the scale would be the "F" section; on the opposite end of the scale would be the "A" section spread out over the same number of points. The rest of the curve is divided the same way. The number of "B" and "D" grades are equal, and the "C" grades are the top of the bell.

As you can see, it all depends on where you land. In some perverse situations you can get an 85 on an exam and still wind up with a "C." (Is that justice?) It is a real drag, and it usually hurts more people than it helps. So do not rely on the curve; try to avoid curve-graded tests. May the gods protect you from the sadistic professor who fancies himself a geometric whiz.

PROJECTS

Projects are the worst. Unless you are an unfulfilled third-grader, a carpenter, or a mechanic, never, ever, ever willingly take a course that has a required project. Projects are time consuming and usually are a vital part of the grade you will receive. Often they are graded for their creativity of concept rather than neatness of construction and innovative content. (This means that the professor is not likely to be awestruck by the fact that you used your keychain to rope off the little driveway made of electrical tape.) What it means is that the clown who makes a film just walking around showing the beauty of Lincoln Center and chatting with the passersby about the additional scope such aesthetic surroundings lends to their lives, is going to get a better grade than you with your model of the Empire State Building, no matter what scale you use. It is known as the "dazzle factor."

So if you have a choice between a paper and a project, do the paper. You can come up with a paper in one night, but have you ever tried designing and building a modern community center for that vacant lot down the street in one evening? Christ, it takes a week to eat enough popsicles to get your building materials! People with projects are always asking for extensions. If you do the paper and get it in on time, the professor will appreciate it, and it will definitely show in the grade.

But if you do fall into a course where you do have to do a project and there is no way out, do it and do it early. It is almost impossible to do an instant project unless it is a piece of modern art—in which case you can commission any three-year-old to do it for a quarter. So build a little of your project every day. Rome was not built in a day, and if you think your little project is going to be easier to build, then you are gravely mistaken. You will not be able to build your little model house, Roman village, computer, airplane, or whatever, any faster. You need at least a day for the glue to dry, or you risk everything when you transport your masterpiece to the eyes of your breathless audience.

4

PAPERS

"When I had to write a paper at the last minute, I used to comfort myself with the thought that by the same time tomorrow it would be all done—typed up and handed in. Sometimes this made me feel weak with relief and I would write away. At other times, it was fatal; I would sit up all night waiting for the paper to appear."

—A Stanford man '75

Knowing you have to write a paper in one day is like knowing you have to build an ark overnight because at the dawn the floods will arrive and sweep away all the dry land. People outside the academic world cannot grasp the intensity of the pressure on a student who has a paper due in less than twenty-four hours. While the real world mutters about keeping perspective, a student, who upon entering academia left behind the normal reality and accepted an entirely new structure of priorities, can drive himself to the brink of a nervous breakdown over a twenty-page political science paper.

But total despair is not necessary (just a little goes a long way toward elevating your heartbeat to the right pitch). Despite what you might have heard, most people do not do their papers in easy little steps. Your high school teachers doubtlessly tried to drum into you that if you do a little every day, then writing a paper will be a painless and rewarding experience. Hah! The only people who write their papers six weeks in advance are chronic worriers, and they tend to suffer for the entire six weeks. It is much easier to suffer a short, intense pain than a long, extended one. (Like the Chinese water torture, people have flipped out because they began worrying about a paper weeks before it was due.)

As for the rewarding aspect of it, there is much greater satisfaction in doing a paper all in one leap than by building a bridge. If you do your paper far in advance, you tend to fuss with it, changing things here and there and without realizing it, you can lose the continuity of thought so that the paper will no longer flow as it should. Most papers are not written in advance. Why do you think you hear typewriters banging away at all hours in the dormitories? Most students pound out their papers in the wee hours of predawn the day they are due. The world is full of procrastinators. The good intentions they had when the paper was assigned two months ago were forgotten, the time has suddenly passed, and now they are frantically typing away. More than one conclusion has been typed while the professor is delivering the day's lecture.

It is difficult to sit down and write a paper when it is not due for another five weeks, but it is harder still to write a paper in ten hours that looks as if you have spent five weeks on it. But you can do it, and you do not have to stage an all-night writing marathon. The problem that most students have is not knowing how to organize their thoughts, and they do not know how to write quickly. The fault lies not with the students, but with the high schools that taught them how to do papers. By simply throwing out everything that you learned in high school and sophisticating your methods, you can turn out a paper in a matter of hours. And you do not have to spend a night in sheer agony only to face a morning of despair.

You read it and we said it—papers are not something to be dreaded. They can be done efficiently, quickly, and with a minimum of pain. You should be able to rely on your papers to pick up your grade average if (God forbid!) you suffer a memory lapse on the final. The proper frame of mind to have when you are writing a paper is to pretend that you are going to be forced to send in a congenital idiot to take the final for you. Therefore you must salvage your grade with this paper.

BUYING PAPERS

Of course, the totally painless way of doing a paper is not to do but to buy it. But here you will resign your fate completely to the person who writes it. If you are not too bright or if you just cannot write papers in general or this paper in particular, then this may be the smartest thing for you to do. But in buying a paper, you must be very careful about whom you buy it from and what you are buying.

Do not bother with the term paper services on your campus unless you want to buy a paper for your kid sister who is in the fifth grade and who does not mind getting a "C." Yes, the "graduate-level" papers from these services *might* pull a "C" from a fifth-grade teacher if she is not very particular. The prices of these paper services are totally outrageous, and the "quality" (pardon the expression) of the work is so abysmal that no matter how desperate you are feeling you should not even consider them. You could write something better in your sleep even with a high fever. The trick these services use is that you cannot see the paper until you have handed over your prayerful pennies. If you saw the paper, you would not give them two cents for it, much less the dollar and up per page they charge. They rely on shafting the desperate.

The best bet in the paper-buying market is a fellow student who took the course or a similar one last semester. Ideally, the paper that you will dig up was not handed in to the same professor. If you cannot manage that, then make sure that it is one that he has not seen for a year or two. (Some professors keep copies of papers handed in to them and others just have ter-rific memories.) When you find the paper, retype the title page. Strain your brain and think up a new title. Then rewrite the introduction and the conclusion. In this way, at least the beginning and the finale will be your own.

If you do get stuck using a more recent paper, then do a minor rewrite of the entire thing, basically to change the style. If you can remember any little in-teresting obscure points of interest that your friend omitted, work them in somehow.

The price that you are going to pay for this work of art is worth discussing; it should be directly propor-tionate to how desperate you are. If you get started early enough on this project, you can possibly even contract a friend to write the damn thing for you at a buck a page or so. (There are students who help send themselves through school by writing papers for the lazies with money.) If you are almost to the wire and find someone who saved his reflection on *Byron's Effect on Russian Literature of the 1820s* and is willing to part with it, you should be able to pick it up for two or three dollars a page (remember nowadays it costs over a dollar just to have a page typed). Most likely you can argue and knock the person down to around forty or fifty dollars for the whole thing. This is if the person got an "A" or "B." If it is a "C" paper, then you are paying just for the outline, the bibliography, and the professor's criticisms. Obviously, you will have to do a rewrite, and the price should be reduced in proportion to the amount of work you are going to have to do.

If someone is willing to give you his old paper, or you manage to talk someone out of a paper, then count yourself among God's chosen.

Finally, when you get this paper from your friend or

business associate, do not let the world know what a great bargain you got. Be quiet about it. These things have a way of getting around. You never know who is going to the professor's office hours every week, cluttering up the halls, and giving him all the latest class gossip. (Those who would tell are aghast at what you did; they are jealous that you may get a better grade undeservedly—but of course your name just slipped out. Needless to say, these people should be taken out and shot.)

A takeoff of this chat may have visions of dollar bills raining down on you as you sell *your* papers. One note: it is not unlikely that you may be able—if it is necessary—to use your old papers again for a different course. Also, the seller—if the transaction is detected—is condemned just as bitterly as the buyer. So think carefully about it.

INSTANT PAPER

Add water and stir well? No, not quite, but almost. You have one day left in which to write a paper, and you cannot reconcile yourself to using your roommate's paper. Or you just could not find anyone to borrow or buy a paper from. Do not despair. Papers can be done in a matter of hours. It is not as good as the original method described in the next section, yet it is definitely safer than buying it or contracting someone else to write it for you. Read on if you are in these narrow straits.

First, forget everything your high school teacher taught you about organizing and writing a paper. You are now a sophisticated artist of the written word. For God's sake, do not rush out and buy index cards (unless of course you like them with your eggs for breakfast).

First pick a subject—preferably one that is widely written about. Now go to the library. Find a pertinent book on your subject. Make sure that this book is either very new (so new that your suddenly profound ideas will seem like a new light on the subject to the professor who has not gotten around to reading it), or that it was published before World War II. The advantage of the older book is that your professor will think that you hold very classical ideas on the subject, or that some very ancient high school teacher made a great impression on your tender mind.

Organizing this paper is very simple. Buy yourself long yellow legal pads (white will do, but yellow is really best) (It may be a roundabout way to point out how vital these pads are, but for what it is worth, Nixon did all his best plotting and planning on them. Ouch!) Condense the pertinent chapters and quotes of this book. Copy the important sentences and paragraphs verbatim. Using bits of your class notes is not out of place here, either. If you can possibly throw in anything you have picked up in other classes, do not hesitate. You are going to be long on theory and facts and very short on personal interpretation in your paper. Now cut up all your little bits of knowledge into separate pieces of paper. Lay them out on the floor in the order that they fit best. Then put your own ideas on little scraps of paper and fit them in where appropriate. You will be surprised—depending on your time factor and how pressured you feel, you will have ideas in direct proportion.

Now start from the top and write transition sentences from idea to idea, quote to quote, from the book. When you have finished, go over the verbatim chunks and put them in your own words, leaving the quotes intact that you want to include from other authors and historical figures. The body of your paper should now look like it is authentically yours.

Finally, write the conclusion and then the introduction. The introduction should lead directly to the conclusion, so if you write the conclusion first, you will know where the paper is going and you can write a wonderful introduction. The conclusion and the introduction should be all your own, and they should follow each other very, very closely. It is sad to report, but true nonetheless, that some professors read only the beginning and the end. If you end by proving what you set out to prove, then you get a good grade. Those who wander to a hazy conclusion can forget it.

Remember: no one says that you have to list your pertinent book in the bibliography (except the morality behind the law). You make up your bibliography from the pertinent book's bibliography. If you throw in enough of your own information from class notes, other classes' notes, and outside bits of knowledge, and you do not copy directly but condense and change phrases, you are not going to come up with that author's work anyway. It will just seem to be a coincidence that you have the same ideas.

All in all, this is plagiarism—not to the absolute letter of the law, but well within the spirit of it. However, if you are desperate it is a very safe way out. Believe it or not, this is much safer than using a paper several years old or buying a paper. If you do it intelligently and take precautions, this is a beautiful way to do a paper fast.

Do not return the pertinent book to the library for a month or so. (Hopefully you have it checked out under a friend's or a roommate's name or it is temporarily missing.)

IF YOU FEEL THAT YOU REALLY MUST: YOUR OWN PAPER

Some people can never reconcile themselves to using shortcuts, and some people really like to write papers. Then again, there are those of us who enjoy playing the game and would not think of diminishing the sport. For people of a similar persuasion, we humbly offer some legal shortcuts.

If you plan to do this paper all by yourself, then set aside some time. There are people who can just pull a ten-page paper magically out of their elite typewriters. They hand them in without the least proofreading and get an "A." If you are not one of these whizzes, set aside the amount of time that you know it will take you to do this paper, and do not forget to allow time to type it if you are not going to compose on the typewriter. We do not mean one day—or it will be a marathon and you will turn out a lousy paper, unless you fudge with our instant-paper method.

First, assemble your bibliography. The fastest way to do this is not to go to your library card files, no matter what your high school teachers told you. Go to the library and use the card files to find a book by an author who has ideas and theories that your particular professor likes. Then look at this author's bibliography and pick out likely books on your subject from his list. This saves you from wasting time on ideas and books that may have been considered tripe from the day they were published. If you are just picking books blind from the card files, this can be a real danger.

The next step is fairly easy. You do not have to read every book you bring home cover-to-cover. You do not even have to look at half of the books that you pick from the first author's bibliography; just copy a few names for your bibliography and take out three or four of the most profound passages, or the ones that have the most material on what you are writing about. These are easily found by looking at the index of the selected books.

When you get home, read each author's introduction and conclusion, and then go through the index and read everything that each has to say on your particular subject. Take notes. And please do not take notes on index cards. This is probably one of the most cumbersome and time-consuming methods that you could possibly use. It cramps your style in more ways than one. Aside from the fact that by the time you finish you will have enough index cards to paper your room with (one NYU student papered his room in early *Peter Stuyvesant in New Amsterdam and His Effect on the History of Manhattan Island* index cards; he had 1,127 eight by fives); the cost of all those cards can break you.

Take your notes on the ever-popular yellow legal pads. Organize by heading each page with the title of the book and the author, and line your notes up below. Before each note, put down what page you got it from (essential if you look at it later and do not quite understand how it got there). You can write in your own ideas by simply putting them down on the same page, but in a different color ink. Do this for each book. By now you should have a fairly concise set of notes, if you remembered to take notes only on what you know you are going to use. Note only the most useful and significant points; skip everything superfluous (almost every student has the problem of too much, not too little). One point to keep in mind; take notes verbatim from your books; if you want to paraphrase, you can, but the quote is there if you should decide you want or need to use it.

After spending all this time taking notes and thinking, you will have formulated your thesis and should be ready to write your conclusion and introduction. Make sure you know what you are going to say and how you are going to "prove" it in the body of the paper. If you write a good paper, the introduction will lead right into the conclusion and the conclusion will merely reiterate what you said in the introduction.

For the body of the paper, separate your notes, and line them up on the floor in the order that they follow best. Write the transition paragraphs. Then do a complete rewrite of the paper so that it flows smoothly. Tack on the introduction and the conclusion and you are finished.

Admittedly this method is just a more streamlined version of the old index-card method. But sixty sheets of legal paper sure as hell beats 1,127 index cards any day.

Last but not least, make sure your paper is neat. A neatly typed, non-typo paper goes a long way toward a good grade. Appearances count. Have decent margins so the professor can write in his comments easily, and make sure the whole thing is well spaced. Professors understandably get impatient (and therefore in the wrong frame of mind) when a paper is sloppy and difficult to read. Also, if you have to stretch the paper to meet the length requirement, work in some little point of interest or an interesting quote. Do NOT wind on for pages and then suddenly turn to a solid conclusion. It will look as if you are stretching. If you are that pushed for length, it is better to widen your margins a bit on all sides, and if it will not be too obvious, triple space instead of double space. Of course, use a pica typeface instead of that teeny-tiny elite. And do have a coherent conclusion; hazy thinking for pages on end to ramble out the conclusion for length has a high irritation value. Remember: the one thing you do not want is an upset professor buzzing through your paper like an angry bee, just waiting to stick it to you.

5

GRADES:
WHAT TO DO
AFTER YOU
HAVE FAILED

Many times in college, you will find that while you were convinced that you were doing superbly, the professor thought that somewhere you missed the proverbial boat. This is not a situation to go over the edge about, especially if you find out during the term. But even if you find out on the eve of transcript mailings, or even *after* you have received your grades, there is no need to panic. In college, NO GRADE IS FINAL. Always keep this in mind when you feel like jumping off the top of a bell tower after getting a depressing "D" or a finishing "F." Writing an eloquent note showing your evident genius before you end your shame is not the way to make the professor aware of how blind he has been. Keep your head together and look intelligently at the alternatives that are available (and we do not mean choosing between joining the Salvation Army or pumping gas at Maxie's Garage). Be aware of the realistic and viable alternatives open to you and then decide in which order you are going to try them. Do not worry; there is a way out. A bad grade is like a maze; it just takes a little time and patience to get out of it.

For example, if you have not reached the end of the term yet and the final, you can usually drop the course, change to taking it pass/fail, or see the professor and find out if he will disregard your terrible midterm grade because of the properly pathetic story that you will manufacture (just remember you will have to prove yourself on the final if you use this ploy). If you have reached the end of the term and cannot get the paper together, you can ask for an extension or an incomplete. (It is always best to ask for these just before the deadline instead of just appearing at the last minute, which only makes you seem careless.) If you are positive that you cannot pass the final, you can ask for an absence or a deferred exam. If you have already received the damning grade, you can go to the chairman of the relevant department and ask to petition the grade, if you feel that you have been dealt with unfairly. (Just remember this is America and the burden of proof is on the accuser. And it is not easy to prove a fox is an ass to other foxes.)

Finally, if you did screw it up all on your own, do not just give up, accept your fate and start asking janitors if they are looking for assistants. What the hell —you have already screwed it up—what can it hurt to go to the professor and beg? (Pride, you say. Bah, humbug we say—are you going to be proud as a ticket taker or are you going to be humble as a manufacturer of the tickets, owner of the theater, star of the play, etc.?) Every professor can send in a grade change to the registrar at any time, up to, including, and sometimes even after the day of graduation. Just one little slip of paper and suddenly everything will be shining again, and you can turn your eyes back to the stars. NO GRADE IS FINAL. Do not despair; all is not lost. You have just lost a skirmish, not the battle, much less the war. Read on and you will soon understand what to do to gain the necessary expertise to survive.

REPEATING THE COURSE

Repeating the course may have been a fun way to spend the summers while you were in high school, but it is not to be considered in college. Do not even let the thought tiptoe across your feverish little mind. If you find yourself failing, get out, drop the course, or take a chance and take an incomplete. But never resignedly sit in that class and get an "F." (Do you want to be the sacrificial lamb offered up to the gods of knowledge on the altar of education? Just decline the honor and run the other way.) Remember: "F's" follow you. They and you for life. They are not wiped from the record █en you get an "A" next time around. Your first █y grade is still there to pull you and your average

█ay need this course for your major, in which █have no choice but to take it again if you fail. █r you find out that you are failing, drop the course, but keep going to class. You are going to need every bit of help you can get. The second—and official —time around with the material, you should do better if you have a complete notebook and some exposure to the subject.

So remember, please, only one grade for every course—and only the best. If you see an "F" looming ahead (and make sure it is an "F" and not a fuzzy "B"), get out while the getting is good.

PERSONALITY CONFLICTS

It may happen that not all your professors will like you no matter how lovable you are. (There is no accounting for some people's taste.) It is hard to think that some forty-year-old bookworm who does not know whether it is raining or not, could suddenly take an intense dislike to the person four seats from the window in the back row. But it does happen. (Maybe you look like the guy who took his first love away, or maybe you look like his first love.) It can happen to you. Even though you will not admit it, you might have precipitated this situation all by yourself by being an opinionated loudmouth, by using your hand as a flag during the lecture, or by criticizing the professor in order to show just how acute you are. These are just a few of the ways to provoke the situation. However, it is possible that you did nothing, that you are com-

pletely innocent and unjustly persecuted. If that is ░
case, all you can do—if you must stay—is sit ti░
Do not despair; you must realize that it happens ░
lot of people. (In all honesty, are there not som░
ple whom you dislike on sight for no conc░
son? If not, then you are either Will Rogers ░
Christ.) Fate has a warped sense of humor, ░
situation usually happens in a small seminar in your
major when it matters most.

There are several things you can do about it to
avoid getting a terrible, unjustified grade. (For despite
the ravings of professorial integrity, if a teacher does
not like a student it is almost impossible for him to be
objective when it is time to grade him.) But it is not
hopeless. (Hopeless is trying to swim the Atlantic to
save air fare. Hopeless is trying to tell your mother
that you are taking the Pill to clear up your flawless
skin. Hopeless is trying to become president if you are
honest, intelligent, sincere, a good administrator, and
poor. These are a few situations that are hopeless. Personality conflicts are not hopeless.) The situation you
are in is merely an extreme difficulty, a bit of bad luck
—that is all. REALLY.

First, if you know that this situation has come about
because of your obvious cavalier attitude toward the
subject at hand or the professor's seriousness, then you
have to change the impression that the professor has of
you. You must start an intensive public-image campaign. Start getting to class on time each and every
meeting. Be very serious in class. Watch those vacant
grins that can be interpreted as sneers. Ask an interested question or two. After two or three weeks of
this, stop by to speak with the professor during his
office hours. Really knock him out with your new attitude, your interest. But take it easy. If you go bursting
into his office on the second day of the campaign ask-

ing some obscure question, and unable to sit still with your intense lust for knowledge, your motives will be suspect. (The Great Pumpkin can sense all insincerity, and so can scholarly professors; they are fiercely protective of their little fields.) Remember: you have to work at this. It must build slowly to the final crescendo where your ardor brightens his sky. It requires acting and it requires patience—but not too much of either.

If you do not have the time or inclination to stage a campaign, you still have a way out. Go to the departmental chairman and ask to be put into a different section of the course. Do not go into your problems with your professor. You hate to change sections, you are learning so much from Professor Creepo, but you have just got to change. You see, you are working your way through school, and you are having terrible scheduling problems, juggling the job and your classes. This other section meets at a far more convenient time, and it really would solve all your problems.

Lay it on him. Do not mention any other problems. As we have said before, professors are thicker than thieves. To oppose one is to oppose all. Overnight you can find yourself branded as a malcontent, a troublemaker, a leper, a character assassin—because no professor is ever, ever wrong in the eyes of his peers. Or at least, no one will ever be able to openly sympathize with you, much less fight on your side. The professor in shining armor went out in the sixties and is not likely to be revived for some time.

If THEY will not let you change sections, the smartest thing to do is to go directly to the registrar and drop the hot potato. No fuss, no mess, just a clean swipe. It is the easy way out.

You can also try to brazen the term out and take your chances on letting Dracula give you a grade. Hopefully, some sort of human kindness will enter his spirit on the day he gives you his grade, but do not count on it. It seems as if past generations have used all the world's allotment of miracles, and all fairy godmothers have either retired or work only for the family. Professors are only too human. Do your damnedest and try to get an "A." Your only hope is that you do just too well to fail or even be given a bottom grade.

It is a hope. But then when you do get a rotten, unjust grade you can try the petition. (Read the petition chapter first.) By all means do this, if it was a graduate student screwing you all term (graduate students, by the way, are renowned for their fangs), it is an especially good idea to petition. But if it was the professor himself, then get your Kleenex out, because banking on a petition against a professor to vindicate you and show for all the world to see just how childish the professor was to have let personalities enter into his grading, is naive and stupid.

Changing your attitude can be done easily. Just be quietly interested; do not crawl and grovel. It can be done with dignity. Changing your section is a bit trickier, but more than worth it if your schedule will not permit a drop. But dropping should not be looked on as quitting. College is a game of self-preservation, and this is just one of the tactics. Use it. Brazening out the affair without any attempt to do anything to relieve the situation can really hurt you, if you have any doubts about getting a fair grade from your archenemy. Look at all your options and decide which is best for you academically and which you can live with. Obviously, you blew it when you picked this professor, so take your time and really think out the situation. Be aware of the results you will get from each alternative. It is your ass. Nobody else cares. If you leave it out, it is bound to be kicked.

PASS/FAIL

The pass/fail is not going to hurt you. Honestly. Ignore the tones of doom with which your adviser refers to pass/fails when he tells you graduate schools consider pass/fails as equivalent to "D"/"F." They do not. (At least not anymore, since the pass/fail has become a widely used mark. Once upon a time, it was true that they were used to be deliberately vague about the grades people like the star halfback got, but such is not the case today.) If you used your pass/fails wisely and sparingly, they are going to do you more good than harm. You see, for one thing, the pass/fail can save your average while giving you the latitude to expand your horizons. It gives you the freedom to take something that you are interested in without jeopardizing your college career. Then, again, the pass/fail can get you through many of the requirements (distribution, that is, since most schools do not allow you to take major courses pass/fail) without having them scar your otherwise perfect parchment. The pass/fail is probably the only academic innovation in the last few years that has actually promoted intellectual curiosity rather than squashed it.

But you do not want too many pass/fails; use them only when you need to. Most universities and colleges limit the number of pass/fails you are allowed to take, either per term or per career, or a combination of both. So you usually do have a limited number of pass/fail options available to you. The option is further limited because you often cannot take any of your major courses pass/fail. This leaves you one-half to two-thirds of your courses to pick from. If you know that you are not going to do too well in English Composition, or in Physics 101, then by all means use the pass/fail.

It is not always wise to sign up right away for your pass/fail at the beginning of the term (unless you must declare your intention to take a course pass/fail when you register). You may be absolutely sure that this is the course that you are going to have deep trouble with, and nine times out of ten you will be right. On the other hand, suppose it is not? What if you discover a talent for it and wind up surpassing your wildest expectations balanced by an unexpected problem in another course? The time to sign up for a course pass/fail is on the last day allowed by law. Wait.

Get as much feedback as you can about how you are doing. Look over the situation and assess it honestly. If you have gotten a "C–" on the midterm in a course and you busted ass studying for it, then you really cannot expect to do much better on the final. Try for your "P;" do not risk a "D." Despite what anyone says, a "P" looks better than a "D." If you do get a "P," the hours will be counted towards graduation. Dropping the course or failing it will get you nothing, and all the time and effort you invested will be for naught.

The most foolish thing you can do with your pass/fails is not to use them, or to save them for a time when you are really in trouble. (Remember the silver your mother never used because she was saving it.) Senior classes are filled with people with 2.4 averages who would never admit that they were in trouble. If their two or three "Cs" and "Ds" had been "Ps," they would be a lot closer to 3.0 averages. Do not be a fool. No graduate school is going to look down on an English major who had the guts to take physics, or an engineering student who braved Russian literature just because the grade is pass/fail. And remember, if you should ace the final, you can always try to rescind the pass/fail and get the "A." But you cannot have a "D" removed for a "P." No way. It is a hard world, and luck belongs to the clever.

THE GOOD EXCUSE (Several Supplied)

There comes a time when every student gets in a situation where he has to lie or die. Lying well is the way out; lying badly is the way to hammer down the nails. Practice lying. Work on your stories. Make them just short of plausible. (Remember: fact is always stranger than fiction.) A bit of the incredible does not hurt. The best lies are slightly fantastic.

But there are some stories that never work, never have and never will, even if they are true. Prime examples:

Dead Relatives: They have heard it and heard it and heard it. It is stupid, it never works, fools nobody, least of all the professor. But if you feel you must use it, then do not go tell the professor yourself. Have a close friend do it. Because if it really were true and you were suddenly bereaved, you would not waste your time going to Psychology 101, being suddenly aware of its insignificance in the scheme of life. Death has a way of putting everything in perspective.

Illness: They have heard this one endlessly, too, so you are going to have to be prepared to offer a valid medical excuse. This is not always the best of ideas because it can lead to checking and then to the dean of students. This dean is a person who has really heard everything, and your upset stomach is going to seem very mild compared to the kid who came down with the bubonic plague last semester. (It seems his laboratory rat was shanghaied off a Chinese Communist submarine.) However, you need not despair if your tale leads you to the dean of students since he is often much more understanding than your professor. He can remain calm because there is nothing for him to take personally—your getting sick and missing the exam is not a rejection of him, but of the professor. (Professors are really very easily hurt, and such sensitive souls bruise easily from imaginary slights.)

Basically, then, the trick is to use excuses that cannot backfire. If no one has really died or you are not sick they can find out. And yes, there are professors who will check. (Paranoids are notoriously suspicious.) But never fear there are excuses that will work. Some of the more eloquent follow:

Excuses That Will Work: (tried and true—only the names have been changed to protect the guilty of Columbia '74.) The key is to use something that is not measurable, something indefinable. The best is the one that can never be proven false: mental anguish (over anything you choose—personal stresses and strains, psychological problems, etc.) You allude to the difficulty but never define it. Even general fatigue works since you are working your way through school. (All those late-night shifts and weekends take their toll.)

One student faced with failing economics went to the dean of students after the professor refused to give her an incomplete. She did not know the work and had not done the homework assignments all term, not because of overwork, but simply because she was far more attracted to going to the movies on Fridays, rather than to the eleven o'clock class. When the dean asked her what had happened, the bullshit flowed:

"You see, I am working my way through school, and one way I get money is engineering at this radio station. And well, with the economics it just got out of hand. Fridays we were supposed to hand in the assignments, but on Thursday nights I have to engineer at a station downtown. I go in about eleven and it is a late night jazz show that goes till six A.M. I tried to do the stuff while the records are on, but you know it is so hard to concentrate. And when I get home, I get time for a short nap, and then it is time for class. I work every night but three. And I am trying to get my thesis done at the same time. And I just do not know what to do about this."

The story worked like a charm, and our story teller got an absence—not even an incomplete.

Another student, on total financial aid in the second semester of her freshman year, was so busy enjoying herself that she did almost nothing. She attended none of her classes, handed in only the barest minimum of work, and did understandably poorly on her finals. The result was that she fell below the average needed to keep her financial aid. Realizing that desperate times call for desperate measures she told the following tale complete with a breakdown of tears:

First she was working a full forty hours a week, since she was not only supporting herself but sent her mother twenty dollars a week. (Her mother was divorced and struggling to raise all this valiant student's younger siblings.) She was having tremendous personal difficulties (deliberately vague and distressing). Not only all this but she was the first and only member of her extended family to go to college, and it was vital that she did not fail. The crux of the problem was that if her grade was not upped she would lose her financial support and have to drop out . . . tears and emotion ended any further explanation. (This clever girl used this at the same time on two professors and got over four points added to her cumulative average that semester.)

Needless to say, the money continued to flow, and she was very careful not to let such a thing happen again. She graduated three years later.

Of course, excuses do not have to be used only for such desperate measures. You can even get out of going to class by utilizing one, and it will give you a chance to cultivate your skill. The heroine of our first story practiced and became a master of the art. One time, this enterprising student wanted to stay home to watch a movie. This meant she would miss her required major seminar which had ten students, where her absence would be noted by a professor who would very much mind. To make matters worse, she had her weekly thesis consultation with the same professor an hour before. She went to the consultation and then went home and watched her movie. The next week, when asked to explain her absence, she revealed that after the consultation she had gone to the park (just to get a breath of fresh air and to collect her thoughts before class) and was . . . mugged. Naturally, she was so upset that she went right home and locked herself in her apartment. The professor was appropriately horrified and asked if our heroine was all right. "Yes," she smiled bravely, "everything is fine now."

You want to tell the truth you say. Okay. Listen, my children, and you shall hear the truthful tale of a naive little dear. A freshman, a literature course, an exam passed out unknown to the student who had skipped the last few weeks of classes. The scene: walking into the classroom to take the final only to be told the exam had been distributed several weeks ago. Panic. After a few moments of fidgety despair in her seat, the student fled back to her room to work on the exam among her books. She finishes writing. She goes in and leaves it on the table. She should have stopped here, but she was insecure about the exam. What did this child do—she believed in honesty. She committed a fatal error. She went to the professor to explain. But she was very upset and could not make coherent noises. The professor misinterpreted the anxiety as guilt over having done no work. When the grades were passed out, the child got a "D" with the comment that her exam was good but since the professor was aware of the circumstances, he felt he could not give her a better grade. (In all good conscience, professors are notoriously conscientious if it will screw you.) This is where the truth will get you. Shut up or lie.

THE INCOMPLETE AND THE DEFERRED EXAM

Beware of incompletes. They come back to haunt you three days before graduation. One Columbia senior staggered to graduation after three weeks of round-the-clock effort to collect all his incomplete credits. He even had had to go back to one of his freshman-year professors to ask the embarrasing question, "Remember me? Here is the paper you asked for four years ago."

Many schools, however, are not so lax about letting incompletes ride; usually you must complete them during first weeks of the next term. This can be a real hassle during Christmas vacation or during registration week. So if you have to take an incomplete, it is to your advantage to get the work done during the last days after your exams; it is worth it even if you have to put off leaving campus for a few days. It is much easier to do a paper or whatever while you are still in gear than to wait over vacation and try to do it all at once when you are out of shape. There has never been a student fulfilling an incomplete who has not said, "Never again, even if it means risking an 'F.' "

A deferred exam is also a form of an incomplete. Here you are excused from the final with the understanding that you will take the exam at a later date, usually at a specific time at the beginning of the next term. The deferred exam must be taken, and it will not be the same one that the rest of the class had. Do not waste your time studying that exam. If the professor is feeling particularly nasty that day, or if he took the deferment personally, the exam can be a crusher calculated to make you crack. A deferred exam is the worst of all possible worlds. For one thing, the class meetings and anything you might have absorbed there are in the misty regions of the dim past; you will have to crack a few books to pass. For another, the professor has time to leisurely pore over your every word, to howl with devilish glee over your every slip, to pounce on your slightest error; for there not sixty exams to grade but one or two and not two days to turn the grades in but weeks stretching endlessly before him. Unless your hands suddenly get cut off the morning of the exam, do not take a deferred exam; the interest you pay for the time is too great. You are warned.

THE EXTENSION

An extension is usually the device by which you get permission to hand in a paper or project during the term, but past the official due date. Occasionally it can be used to turn in the final paper late, but for the most part it is a during-the-term lifesaver. It is only politic to meet the extended deadline your professor gives you; after all he was nice enough to give you the extension in the first place.

Be appropriately grateful when you turn it in either with a note inside or some such motion of propriety. Spend a little extra time on the paper, especially on its physical appearance. Smudged pages and blatant typos can be forgiven if you rushed to turn a paper in on time, but if you have been given several extra weeks to put the paper together, it does not go down so well. This is particularly true since the most common excuse given for needing an extension is that you are pressed for time and have done all the preliminary work but just have not been able to pull the paper together in a final form.

Ask for your extension at least a week before the due date. It shows that you plan ahead and are a conscientious student. If you know that you will not or do not want to do an instant paper for the course, then get an extension. One point to remember: never plead that you have three papers due all on the same day. Although you will probably get the extension, the professor is going to feel resentful that he is the one who has to wait. After all, is not his course the most important one you are taking? Professors obviously have a different set of priorities. But extensions are usually readily granted with little explanation as to the whys and wherefores, and they can do a beautiful job of distributing the burden. The main point is not to abuse the Archangel extension.

THE ABSENCE

You may have never heard of "the absence." It is a secret that a few schools keep deeply hidden in the catalog. The absence is a subtle form of the drop used to protect only the more promising students. You took the course, but you were absent from the final. Your transcript records the course but no grade. However, it is not an incomplete; there is no work to make up and no deferred exam over your head. The absence falls somewhere in the nebulous ground between the drop and the audit.

Getting an absence is great if you can swing it. The course shows on your transcript, and it shows that you tried. Yet when you failed you were not penalized by a "withdrew passing" or a "withdrew failing." Anything this good obviously does not come easily. Getting an absence is rather like walking a tightrope; if you fall, it is into the web of incompletes. Getting an absence takes some fancy footwork and in most colleges a consultation with the dean of students. The time to try for this is after you realize that the drop has passed you by, and that even if you do manage to get a deferred exam, you probably will not pass.

However, this divine dispensation is handed out only to promising scholars, so you need to have at least a "B" average, hopefully better. Do not bother showing your face if you do not. The dean of students is going to see that once again you were messing around and just could not get yourself together for the final; sympathy will not be forthcoming. The only way to get the absence is to stand on your good record which is woefully and distressingly in danger of being destroyed. Go to the dean of students after the drop date has passed, and fall apart right then and there. Tell the dean that you cannot have this on your record, you will not be able to get into any graduate school with such a flaw, you just do not know how this got away from you so fast, and now . . . and now . . . now it is all over. Your academic life is over. You are being pulled out to sea.

Look as though you are going over the edge, making a few remarks about giving it all up and going home are not out of place at this time. But try not to go overboard and start babbling about going to beauty schools or Driving Training Institute. You are far too upset to think about the future; your life is academic. Is there anything else?

It takes chutzpah and a decent average, but an absence is a marvelous deal if you can get it. Try it—what the hell. . . .

DROPPING THE COURSE

Drop a course only as a last resort. In some schools you can drop a course almost anytime up to the final without going to the professor and getting a permission slip. All you do is get a little slip from the registrar, fill it out, and prestochangeo you are no longer a member of the class. (Make sure the registrar records the little slips, since little slips slip into little cracks and disappear. You will be in for a rude shock if you get your grades only to find out that somehow your drop was never recorded and in its place sits a beaming "F.") In other schools, filing a drop can be a little more complicated. It may be that you can drop a course at almost any time, but your transcript will not record a drop but rather a "withdrew passing" or "withdrew failing." Most people who receive a "withdrew failing" on their record are the ones who simply did not bother going to the professor first; they just filed with the registrar. The professor received a memo requesting the status of the disappearing student (and as we said they are easily hurt and strike back by reporting "withdrew failing"). If you jump the gun on the registrar and visit the professor and ease the blow of the memo, you can in most cases escape nearly unscathed with a "withdrew passing." It is harder for a professor to say no to you (a person who is properly regretful and sad that he underestimated his course load and is swamped by work and so sorry that she/he has not been doing well and would never think of staying in a course without the time to give its deserved attention) than it is for him to write a failing notice for an anonymous person whom he has never met.

If you find yourself floundering and cannot possibly sign up for a pass/fail, then by all means drop. Graduate schools have no use for the heroes who try to brave out the finals only to end up losers. They love winners. By dropping, you are a winner; even though it may be hard for you to resign yourself to your failure, it is probably easier than trying to comprehend the intricacies of Biology I. If you find that you still have to take this course for your major, drop it but do not stop going. Audit the course unofficially for the rest of the term; you will be that much ahead of the people who are going to come into it cold the next term.

Finally, if you get past the drop date before you can bring yourself to admit that you are in deep trouble or before you are aware of the undertow, do not despair. You can still drop the course. In the process you may have to see every adviser, half the department involved, and the dean of students, and get a million and a half approvals but it can be done. There are very few hard and fast laws; most rules are guidelines and you can always be an exception. Just as no grade is final, no deadline is final.

FOR THE FEARLESS ONLY: THE PETITION

No grade is ever final. If you feel that because of personality conflict with your professor or because of a graduate student's prejudices, or just through oversight, you got a grade that you did not deserve, you can petition this grade to a committee of the faculty from the offending department. (If the grade was from a graduate assistant, you can first try petitioning the professor.)

Petitioning is a bit difficult and can get very sticky. You gather up all your evidence, your tests and papers, and saunter in to the head of the department. Tell him your story and give him the papers. Then he will give you a date when you will get your decision. A small committee of the department concerned will meet, look over your papers, and see if your claims are justified. If they are not, God help you. Professors are thicker than thieves, and you will never be able to take another course in that department as long as they live or their memory lingers on.

However, if you are justified and the committee decides in your favor (however slight the chance may be, Lady Luck sometimes sets down at a college to rest on her way to Las Vegas), your grade will be changed and you will be triumphant. But you will still be taking your life into your hands if you take a course in that department again. And it goes without saying that you are suicidal if you sign up for a course with the same professor. (Nobody likes having salt rubbed into his wounds.) May we suggest that you do not try it in your major department unless you are absolutely unquestionably in the right—that is, unless you have written papers that make Tolstoy and Joyce look like hack writers or scientific reports that have Russian spies trying to get you to defect or kidnapping you. Unless there is no question that the rest of the department faculty are going to be as outraged and bewildered as you are, FORGET IT.

TRANSFERRING

If you know that you hate your school after the first term and feel that you did badly because of it, transferring to another college is a viable alternative. But transferring is difficult, you can lose credit and time, and there is always the possibility that you will hate your second school more than you hated the first. (Or that it is not so much the school as the prevailing system that is killing you.)

If you are considering transferring, you should sit down and ask yourself some hard questions.

Do you want to transfer because you are miserable and miss your family and friends? (Unless you are going to be a lifetime boarder at your parents' home, you should forget transferring simply because you miss Mommy and Daddy. Unless you transfer to your local community college, you are going to be just as unhappy anywhere else.)

Are you unhappy about the grades you received and think that you are not going to be able to do much better during the next three years? (Most freshmen feel this way, and transferring is rarely the solution. Unless you are going to quit altogether, then you are going to have to face these problems sometimes. More work, more plotting, and more study is the answer, not a new school—unless you go to Retard U.)

Are you considering the move because you cannot stand the people? (It may sound trite, but people are people. It is true that sometimes a country boy cannot adjust to the city or that a city boy cannot take the pace of the country, but in most cases all you need is a change of roommates or dorms to lighten up the atmosphere. Jocks tend to live more happily with jocks, grinds with grinds, Greeks with Greeks, etc., etc. You can gain a whole different perspective on the campus simply by moving and finding your own kind of people. Look for them. It takes all kinds, and every campus has a few of everything. It will be just as hard to find them at the next campus as it will be to look around you—maybe even harder.)

Not every reason is a bad reason to transfer. You should transfer if you find that you do want to specialize in something that your current scool does not offer. If you can get a B.A. in the specialty, do not waste time taking something vaguely related and relying on graduate school to understand and train you. Get with it. Transfer.

If you are desperately unhappy with the place and the people, and you have given the school a fair chance (a year, not one term), then leave. Just the change of scene will probably do you good. Finally, there are those who find that their college is not really stimulating enough. (The way these people got to said college is always a saga in the first place, usually involving high school problems, terrible SATs, and no recommendations. Having rectified past problems, brought up their grades, and matured a great deal, they are now ready to move on.) If you feel this way and feel that you are now ready for a heavyweight school, then start sending for catalogs.

If you have decided to transfer, have a plan and a definite school that you want to get into. The most important thing you want to take with you when you leave are your earned credits. So keep your interests in astrology, science fiction, and the design of medieval villages at bay until your transfer is completed. Take a Shakespeare course instead. Register for courses that will be easily transferable. When you attempt to transfer credits from more specialized courses, you may not be able to get them because your new school might not have anyone around to evaluate the work you did or the content of the course. The more basic the course, the less trouble you have trying to transfer the credit.

The grades you get are also very important, and it is essential that you get at least a "C" in everything. Many schools will not even consider accepting transfers unless they have a "B" average. But you should be more concerned about transferring the credits, and nothing below a "C" is transferable in most cases (country club universities, and state factories excepted, of course). Your "D" and "F" grades are dead losses.

Get to know at least one of your professors well, and make sure that he knows the depth of your misery and academic anguish. He should be aware that you would be much happier at Blank U. You will have to get a recommendation from a professor as well as the approval of the registrar. At this point, your high school recommendation (as well as your SATs and grades) are almost obsolete in regard to your acceptance or rejection, which may be to your detriment or to your benefit.

If you are an unhappy prelaw or an unhappy premed who plans to continue majoring for law or medicine, but at a different school, do not say so on your application. Most schools have more than enough of these people already. They are not looking for people to fill up the window ledges in these classes. Look at the list of the majors offered at your prospective college. List your major as Greek history, cinematography, Serbo-Croatian, etc. Look for the department that is most forgotten and neglected, that needs people to survive. After you get into your school, you FIND YOURSELF and register in the department of your choice.

Transferring is not something to decide on on the spur of the moment. Give your school a chance and see if the problem does not really lie with you and not the institution. But if you decide to go, then go. Do not waste time. Pick a new school and apply. Work out the most opportune time, make your plans with an adviser there, and beat feet.

But just remember: you don't know what you've got until it is gone, and the grass is always greener. Then again, it is always darkest just before the dawn. Nobody knows the trouble you have seen, so just make sure that it is really what you want.

SLEEPING WITH THE PROFESSOR

This was the classic way for the wayward—and gorgeous—student man or woman, to get a good grade in the days of the charleston, raccoon coats, and hip flasks. These were also the days of the *Titanic,* the *Hindenburg* and the first atomic bomb. You will be courting the same kind of disaster if you try this tactic today.

Honesty is the byword today in sexual relationships, and if you do sleep with the professor, he or she is likely to bend over backward to grade you fairly. Chances are that you will be graded more harshly than the other students. Instead of the "A" and a grateful tweak on all four cheeks, you are going to get the "D" you deserve.

Sleeping with the professor today will probably get you screwed in more ways than one.

AIN'T TOO PROUD TO BEG

You should not be, if you're really up against the wall. Begging should be the last resort, but misplaced pride is stupidity. Since the draft ended, there are not many life-or-death bits going down with much success any-more. But this has not killed all the odds in begging; it has merely refined begging to an art.

And begging is an art. (Ask any professional beggar in the world.) For your purposes, begging is centered around learning how to give the impression of deep psychological problems. You never have to nor should you go into the specifics. (The unknown is far more powerful than the known.) Allude, insinuate anything, but the fact that you just were not prepared for the exam or that you could not be bothered to do the paper. The vaguer the problems, the better. The best ploy today is the overpowering mental anguish you are suffering because of the way your personal life is going. It is so insufferable that it renders you incapable of thinking, working, etc. You are simply too upset to think straight. Do not worry—after flunking you will be in the proper frame of mind to pull this off. And do not be afraid to let the tears flow. For the men, the breaking voice is an excellent technique because it will seem as if you are going to plunge over the edge at any moment. Do not go in like this; build to it slowly during the course of your private chat with the professor. Do not fall apart visibly. You should hold it just under the surface. (The tension the professor will feel will be enormous; he will be on edge waiting for the explosion.) Be apologetic. Show that you really care that your academic career is suffering during your time of trial and tribulation. In fact, your inability to fulfill your school obligations is the final indignity—the proverbial straw.

As we said and you can now see, begging is an art. Your begging should show the proper combination of dignity, anguish, responsibility, and a tinge of sadness. You are one of the age's noble heroes who is having difficulties adjusting and coping. Become your own martyr. Do not say this to a mirror or practice before you go; you will blow it. Try thinking of flunking out and becoming a waitress or a mechanic. Then tell yourself that begging is an art and you are one of its greatest artists. You should go down in history.

As we say, misplaced pride is stupidity. Break a leg.

6

SOPHOMORE
YEAR

All in all, sophomore year has to be the best year that you will spend in college. You are still taking the very basic courses with one or two exceptions, so that you should not have too much trouble academically. You have also managed to gain some perspective and academic street knowledge so that you should be able to cope with any problem that arises. But best of all, you know your way around, and you have friends and one hell of a social life going (all but inevitable unless you were a monk during your freshman year).

There is not much to say about sophomore year. You will have problems (how dull would life be without them). You will have to avoid declaring a major and to make sure that your money does not run out (and we will tell you how to avoid this). But other than that, HAVE A GREAT TIME.

DECLARING YOUR MAJOR AND HOW TO AVOID IT

The school will try to pressure you into declaring a major at some point during your sophomore year. Unless you are ABSOLUTELY, DEFINITELY AND DECIDEDLY SURE that you know what you want, skip it. Do not do it. Do not tell anytbody anything (at least not officially). The minute you sign up for a particular major, you are going to be stuck in the workings of bureaucracy.

Theoretically, it is very easy to change your major if you have been taking very general courses and concentrating only slightly in one area. In practice, though, it is a very different ball game. At the end of sophomore year, you will not have enough credits in anything to slow you down irreparably and put you on the five-year plan if you should decide that you want out. If you have not registered a major, you can just change your mind and proceed. (It *is possible* to fulfill all major requirements in two years.) However, if you have registered a major, it is not that easy. You must change your major officially. To do this, you must get the signature of the department you are leaving, the okay of the one you are going to, the registrar's stamp of approval, and in some cases, have an interview with the dean of students. It is a big hassle, and it never ends right there. It goes on, and on, and on, and on. The registrar forgets that he must tell the Big Computer that you are in a new major, and you find that you have been shut out of vital classes with the rest of the nonmajors. You are left off class and department lists. You have trouble getting a new major adviser. It is a bitch.

In essence, it is best not to bother until you are 101 percent sure. Sophomore year is not the year when you are sure—about anything. Changing your major at the end of junior year and losing priceless credits is really something to worry about. This sort of nonsense that you must go through just because you filled in a blank with any old thing that popped into your head in sophomore year is fortunately avoidable. Just leave the blank blank. No matter what sort of tap dance the registrar does on your head, if you do not want to declare a major, then don't. He just wants to keep his class files complete, the classes balanced, and the Big Computer happy. He does not give a flying fuck about the hell it can cause you later.

FINANCIAL AID

There are so many places to get money to keep your academic career afloat that it would be impossible to list them all, or even discuss the hundreds of major sources. The best we can do is to offer some pointers on where to look for the money and how to keep it in your hot little hands after you get it.

Grants and Scholarships

It is estimated by some experts that there is literally millions of dollars available in various funds for higher education that goes untapped each year. Much of this money is sitting in trust funds throughout the country. Do you remember the rah-rah types that we said would leave their estates to Alma Mater? Well, they did just that, but by the time they actually got around to making out their wills they had become senile. The result of which was they left a scholarship available to any student who is five-foot-two, wears a size forty shoe and comes from Grasstown, Iowa. The problem is that any five-foot-two student from Grasstown, Iowa who happens to wear a size forty shoe may be totally unaware of this potential source of funding. This problem has been solved to some degree by new businesses such as Scholarship Search (1775 Broadway, Room 627A, New York, New York 10019). These services send you an exhaustive questionnaire asking anything that might help qualify you for aid. Then, for the minuscule price of $39 (or thereabouts), they will run your qualifications through their super-duper computer (which knows all about these hidden little pockets of wealth). This particular outfit has a data bank listing some 250,000 different sources of financial aid available today. They guarantee that they will find you at least five sources for you to apply to and exploit. If they cannot come up with five, they will not only give you your money back, but they will give you whatever they were able to come up with for you, free of charge. (The testing services run a similar search operation. Contact either your SAT or ACT center.)

You still have to go through application procedures for the money and prove your worthiness, but Scholarship Search claims that one out of every two students it handles receives aid from one or more of the sources it puts the student in touch with. You can find out more about these organizations by calling your local public library or your college's financial aid office. (Also double-check with the Better Business Bureau before sending them any money.) It is definitely worth trying, as you can see. Who knows, the fact that your grandmother had a mole on her left index finger and shopped at Bartley's Bakery on Wednesdays at 3:31 and bought a loaf of bread, half-sliced, may be just the requirement necessary to get you $1,000 a year for four years.

The problem for many students on financial aid is continuity. Unfortunately, many of the tiny graduation prizes and grants of $250 (from the little old ladies of The Society of the Lending of Comforts to the Sick of Parsippany) tend to run out at the end of your freshman year. This leaves you high and dry. You have to find other funds to keep your career going—or take the next train home. The financial aid office at your college can often be a great help to you if you have suddenly found your well has run dry. They can help locate new funds, if you give them a chance, and let them know that unless they help you, you will disappear. But you have to keep them advised of all changes in your financial position to expect them to help you. They do not take kindly to people walking in and announcing they are broke the second week of the term. Or people who wait until literally they do not have a full dollar to their name. It takes time to put together a financial boat; you risk drowning if you wait until the last minute.

If you are doing well, they will tell you which of the university prizes and grants you are eligible for. Several of these are given to promising students every year, and they can range from a fifty-dollar book award to a year's tuition. You have to apply for these, so check them out and see if you could possibly get them. It never hurts to try.

Occasionally you can win a prize for poetry or journalism. Some schools give free dorm rooms to the editor of the campus paper, the president of the student body, or the captain of the football team (in the Midwest, the whole football team, which almost makes it worth joining). It is also common practice to give floor counselors a free room (and often a small monthly allowance. However, the ordinary mortal will have to keep his eyes and ears open and make frequent trips to the registrar and the financial aid office to find out about the university opportunities. Obviously there is a limited number of places open for *an* editor or *a* captain, etc.

Loans

Aside from grants and scholarships, there are institutions that will loan you the money to expand your horizons. Among the most popular are the banks (which loan out money under the Federally Insured Student Loan program). The obvious drawback to taking out loans is that you have to pay them back—often with interest—and if you are going to an expensive school, it can run into the tens of thousands of dollars. Despite all, however, it is inevitable if you are a total financial-aid student that you will be allowed to escape unscathed by loans. Most schools today use a packaging system whereby they calculate your needs and then put together a package to meet it. This package is composed of your grants, scholarships, and suggested loans. Therefore the money you can come up with from various trust funds, etc., will go a long way toward lightening your borrowed burdens.

If you are an out-of-state student attending a state university, look into the state residency requirements. By having yourself declared an "emancipated minor" and then fulfilling the state residency requirements, you can become eligible to pay the state resident tuition fees, which are usually substantially lower than the out-of-state fees. The financial aid office or the local board of elections can fill you in on the specific residency requirements.

Once you get many grants and scholarships, you have them for the full four years. Loans, however, usually need to be renewed yearly (unless you took out a large loan and are merely using it in pieces). For the most part, to keep your financial aid (especially school grants) you must maintain a specific point average. Maintain it. If you do not, you may lose some or all of your funding and may be forced to drop out. (Often if your grade average falls below the required level, you are put on suspension until it is brought back up. To a student on total financial aid, a temporary cutoff might as well be permanent. Either way it spells the end.)

Another point to remember—especially important if your sources of income are scattered—is to know the deadlines for renewal applications where necessary. Almost every school requires the standard PCFS (Parents Confidential Financial Statement) filed through the test services. It is usually due at the beginning of the year, so make sure you get it to your parents in plenty of time for them to fill it out and send it in. (And keep after them about it—parents have been known to forget.) Aside from doing your part, you have to make sure that the registrar is doing his part. It's wise to be sure it doesn't take him three months to mail out your grades and the "student in good standing" form. (It may seem a simple procedure to you for him to look at your record and put a little "x" in the appropriate box, but to him it could take ages.) Check. It is your neck, not his.

Work-study

Last—and sometimes least—are the work-study programs available for students on most campuses to provide them with eating money. You have to prove your financial need to get on the work-study program, a federal program whereby the U.S. Government covers part of your wage, if not all of it. This is usually accomplished by the PCFS. If you know that you will need money for food and other necessities as soon as you get to campus in your freshman year, it's wise to sign up for it. The work-study program "guarantees" that the school will find you a job. For a freshman coming into a new town and making so many other adjustments, having to worry about finding a job can be a tremendous burden. Let the school worry. They will find you a job in the dining halls, or doing clerical work for a professor, or shelving in the library. It rarely pays more than the minimum wage, and you are often allowed only so many hours a week. But if you count the hours of mental anguish that you will possibly go through before you find a job in a strange town during the first week of registration and classes, it is worth it. Later, when you know your schedule and you know your way around, you will be able to look for a more lucrative position.

You just have to keep your eyes and ears open. Try for all sources that might possibly be open to you. There are ways—lots of them—to make money on campus, and many people who will be willing to lend you or give you money. All you have to do is find them. They do not hand it out on the street corners—you have to search them out. Get snooping—the sooner the better.

7

JUNIOR YEAR

This is when you should let it all go, and DECLARE YOUR MAJOR! If you have been smart, you have managed to avoid all the registrar's little notes reminding you to declare a major . (These started to appear in your mailbox like clockwork after the first week of your freshman year.) But now the time has come to declare yourself an aspiring whatever. It should be the first thing you do after getting your key to the dorm. If you hold out any longer, you are going to find that it is working against you, rather than for you as in the past. You see, junior year is the year of the required seminars for majors—and you cannot get into them unless you are a declared major. That does not mean climbing on top of the campus statue and shouting to the world your secret desire to be a major poet; it means filling out the appropriate little slip and dropping it into the appropriate little hole.

There is no way to get around this. To get your degree, you have to have majored in something—they will not give you a degree in Basic Course Taking—and to major in something, you have to have taken several small classes for majors. These small, speedy classes are seminars, naturally (they feel you are not going to get the necessary expertise in nice, slow lecture classes). Seminars are hard—there is no denying that —but you do learn a great deal. The biggest problem involved for you now in taking a seminar (after all, you now have the street tactics to use on the academic side of it), is getting into the one that you want. You will always be able to get into *a* section of the seminar to fulfill the requirement, but—and it is a big but—by now everyone knows who the good professors are and who the windbags are. You will really have to be on your toes to get the section and the professor you want.

Junior year is also the time to start looking forward and planning your strategy for getting into graduate school. It is not too early. You should get to work getting your professor recommendations together and exploring the opportunities that are available to you. You should also pull out the old transcript and go over it with a magnifying glass.

You have probably managed to fulfill most of your distribution requirements. Get a clear picture of exactly what you are going to have to pick up. Make sure that your distribution credits are properly spread out. (Usually you can only take so many credits from a set of departments outside your declared major.) By now you should know every gut on campus. Take one or more of these to clean up your distribution problem. Unless you are applying for your own university's graduate school (and often even if you are), no one is going to know that The Architect in Society was the biggest gut on campus. Or that after the midterm you had the option of taking your midterm grade as a final grade or of taking the exam final to raise your "C+." Or that four people out of three hundred got "Cs." If you pack your program with one or more of these guts, it will look as though you hit an intellectual peak, took a tremendous course load, and aced it. Only you (and everyone else on your campus) will know the truth. No one else. (Except maybe your shadow.)

You have to be on your toes junior year to stay ahead. Do not think because you know the system that it is all going to be gravy from here on in. It is not. The minute you let your guard down, it will all fall down. Just make sure that none of it falls on you. While junior year is one of the most relaxing, if you play your courses right, it can also be the most profitable.

REQUIRED SEMINARS

Every major has its required seminars. These are usually sectional and very competitive. But the competition starts when people sign up for professors. By the time you are a junior, you have had a few courses in your department, and you know which professors you like and which like you. So does everyone else. You are obviously going to have a strong preference as to the professor you want. Unfortunately, the difficulty arises when everyone else wants the same professor and there are only ten or so spots in his class. The only thing you can do here is to be ruthless. Do not forget it.

Seminars are rarely chosen for their content. It is the professor that is at stake, because this is the person who will usually be writing your recommendations. Everyone wants to get into the nice professor's class and impress him. If you are not quick, you will not only lose out on your first, but second, third, and fourth choice, and wind up with the miserable hag whom everyone hates and who hates everyone.

Judge the professors very carefully in this matter. Some professors consider attendance mandatory in a major seminar. No matter how good your thesis or paper may be, if they have not seen your bright eyes every class, you are not going to make it. Then there are others who are very philosophical and realize that there is a lot going on in your life at this particular moment, and will judge you only on your concrete written work. Obviously if you are not a class-attender you cannot expect the professor to give you a glowing recommendation. He may not even recognize you when you come to give him the form.

Required seminars are part of your major. The grade you get in them is a vital part of your transcript since they are the closest to graduate work you will get unless you take a class in the graduate department. In the liberal arts, seminars will be intensive reading courses, with a lot of heavy paperwork, and in most cases, they will ask for a thesis or a thesis-length paper at the end of the term. In the hard sciences, seminars will be intensive study courses; papers are not normally required, but sometimes rather than an exam, there will be a comprehensive exam that will entail your remembering a great deal of the basic work you covered since your freshman year. Study!

It is precisely because these courses are so important that you should fight like hell to get into a course that interests you, taught by a professor that you like, and who returns the compliment.

So get in there, climb over bodies, beat people senseless, and give it one for the old Gipper.

THE MAJOR ADVISER

When you do finally ("Thank God," says the registrar) sign up for a major, the department will assign you an adviser. Schools differ so widely in the way major advisers are chosen that it is difficult to generalize. In the best of all possible worlds, you are asked if you have a preference. In others, the majors are simply divided up alphabetically (or schedulewise, or by height, however). Regardless of how you are assigned, your major adviser will be on the faculty of your department. He or she will know and should be able to explain just what your major requirements entail. This is the person who will help you work out the technicalities of your schedule (that is, which of your courses taken at other colleges in the university will count as major credits). While the major adviser is on the faculty and therefore part of THEM, he is the academic prototype of the guardian angel. You are a fool if you do not take advantage of the situation.

Your major adviser is an excellent source of information concerning the courses in your department (but do not expect him to give you an honest evaluation of Professor Deadleaf's teaching skill). He can tell you what the courses are really about—aside from the glowing catalog description. The romantics course can sound great, but it might not have been designed with the upperclassman major in mind. The major adviser can warn you before you find yourself stuck in a class of freshmen reading Keats's "Ode to a Nightingale." (Another less pleasant function of your major adviser is to make sure that you have a rounded program and do not take eighteenth-century literature courses exclusively for your final two years. This means that he will probably nudge you at the beginning of each term to take a course or two that do not exactly spark the old learning fires. But such is life. After all, your degree will say "English literature"; it is in graduate school that you can narrow it down to 1743.) So do go and at least talk to the man.

Aside from helping you out with your undergraduate problems, your major adviser is the best person to run to for advice on graduate schools and getting into them. He can usually tell you which graduate schools are the best for your particular interests and which would be most likely to accept you. (And if you are very good kid and become his friend, he might write a personal letter to his graduate adviser at his old school and mention that you—a wonderful student with tons of Potential—just happen to be applying and that they should grab while the grabbing's good. Would he be on guard and watch for your application and give it a boost?)

You can also count on your major adviser to write a

decent recommendation for you if you have made an effort to see him a few times during your career. Do not turn your back on him; he can be your one good friend on the faculty, and many times you may well need a friend. Do not be like the Yale senior who needed to have his adviser sign his graduation card, only to have the professor cut him short, saying, "I have no more room for any more freshman advisees."

SABOTAGE

It happens—and it can happen to you. As you get into the major class levels of the sciences, you will find that the atmosphere in your classes has cooled considerably. Every student is out for himself. If this includes sabotaging another struggling student's experiment, stealing his notebook, giving him the wrong test corrections, or cutting the vital pages from an important book that only the reserve library has, then that is what he will resort to—and he is willing to do it.

So do not bother asking anyone else for help with an especially difficult problem in organic chemistry or advanced physiology. The answers (if you get any) will not be reliable. Take some time off, or take your books with you and go study in the hall while you wait for a minute of the professor's time. It may take an hour or two, but it will save you a frustrating and wasted night to unravel the wrong answer that Joe Genius (alias Phony Tony) gave you.

If you do find that your experiment has been tampered with, or that your notebooks have mysteriously disappeared from the john ledge, or that the pages are missing from that special library book, go directly to the professor and tell him. There will not be a last-minute reprieve when the guilty party breaks down and, leaning on the professor's comforting shoulder, confesses. Most likely the creep will go off happily with a great grade, and you will get shafted royally. So do not play the fool. Tell the professor. Lucas Tanner and Mr. Novak types do not exist in college. The professor is not going to instinctively know that something fishy is going on unless someone tells him. One look at your downcast face will not flood him with empathetic insight; for all he knows, you may simply be suffering from a hangover.

Ask for an extension or a deferred exam. And when you do redo your experiment, or track down a whole copy of the book, or rewrite your notebooks, guard the damn thing. (Lightning can strike twice, you know.)

HOW TO FIGHT WITH THE REGISTRAR AND WIN

By the time you are a junior, you will have developed your own methods of dealing with the registrar and his inevitable inefficiency and obstinancy. You may have decided to use the soft-sell approach, and since it has proven successful in the past, to stick with it. However, if you have had very little success in the past, it is time to change your tune. New tactics are called for. The registrar, you must understand, is Satan's devout servant, and his efficiency, or lack thereof, is going to become very important to you when you start filing recommendations and sending out applications. It is vital that you get the courses you need and that you get your transcripts out on time. If the registrar fails you, you are going to be up the creek without the proverbial paddle.

After spending anywhere from four to eleven thousand dollars to put yourself through school up to this point, you do have some rights (although having them and getting them are two different matters). If the registrar is not going to help you, you have the right to mention that you have invested so much money and that you do not have to put up with this nonsense.

Many of the registration departments across the country adopt a "next to God almighty" attitude about their work and forget that they are supposed to be working for (and not against) the student. (They tend to get confused and to think that their function is to make life as difficult as possible for him.) If they are messing around with your courses or wreaking havoc with your applications by forgetting to stamp them and send them out quickly, do not suffer in silence. Say something—and that does not mean bend your roommate's ear. Writing a letter to the campus newspaper can be very effective, but you can start a grassroots groundswell that can spell danger and trouble. (Your problem will be solved immediately, but God help you the next time around!) The best way to handle this situation is to take your problem to someone higher up. First try your adviser. If that does not help, go to the dean of students, and keep on going if you have to end up in the president's office.

If you are having problems that even the dean of students is having difficulty solving (and the president is inaccessible) then go to the dean of the college. This will really put the heat on. Try not to create a ruckus, though, unless they have really botched it up. (While your would-be graduate school will be sympathetic, they will only be able to offer you sincere regrets if the registrar did not get your complete transcript to them on time.) So if the registrar has really cost you, do not hesitate to make your move. Even the stockholder with one share can speak up at the General Motors annual meeting. After all the time and money and effort you have invested, you certainly have the right to speak up and COMPLAIN.

8

SENIOR YEAR

You have almost made it. You have rounded the bend and are now in the home stretch. You are so close to winning that you can smell the roses. But you cannot afford to let yourself be overcome by euphoria. Now you know how to schedule yourself and how to get around most of the problems that seemed staggering to you as a freshman. Do not worry—you are ready for the formerly insurmountable problems of senior year. However, beware; your most important trials are yet to come. This is where the competition really becomes hot.

For this is the year of the thesis, the law boards, the med boards, and your applications to graduate schools. This is also the year of the final round with the registrar. It can be a bitch. The hassles you will run into with these people will convince you they are the devil's own and make you wonder why you ever bothered at all. It is incredibly depressing to realize that you broke your back for years only to have it monkey-wrenched by some asshole's losing a piece of paper. The registrar has been known to lose entire records, to have forgotten to have entered grades in the first place or the changes in them, and to have let certain professors go for four years without having handed in your grade in a vital course, etc., etc. ad infinitum. So be prepared for graduation woes. By the time you actually get on line to get your piece of paper, you will really deserve it. But the registrar will have his final laugh; your name will probably be spelled wrong.

HOW TO WRITE YOUR THESIS QUICKLY

Okay, you know that you had over fifteen weeks to write this damn thing. Now you are down to the last two weeks or maybe even one. Only seven days left:

one hundred sixty-eight hours; ten thousand eighty minutes; six hundred four thousand, eight hundred seconds. A veritable lifetime, you say. However, if you have less than this lifetime left, you had better not waste precious seconds reading this section, but skip over it to the next one.

For those of you still with us, sit down and take a deep breath and relax—it may be your last chance for ages. If you have over a week left, then you are on easy street; you still have plenty of time left. Think about what you are going to write. Reconcile yourself to the fact that you are not going to write the most innovative, startling paper on something that has gone neglected for decades. Skip it. You are in ye olde desperate straits. You must get yourself together and write about something you know (hopefully well—yes, if you think about it, you know a lot about several things). You have time to do some research and a little rewriting, but there is no time to fool around. Move it! (What are you doing sitting in that chair relaxing—do you realize you have just squandered three hundred seconds? Move!)

In an earlier chapter, we told you how to do a quickie paper. This is going to be the same thing, only a bit longer. However, you can follow the same basic mode of attack. Get several books pertaining *directly* to your chosen subject. (What do you mean you have not decided exactly what your theme is to be . . . the seconds are ticking away. Decide and write it down. Make it specific. Short and sweet. And do not deviate from it—ever.) Select one book as "The Book." (This is the book that is going to decorate your bookshelf for as long as you are going to be able to renew it, or indefinitely, if it just appeared in your room without being checked out.) You get your basic facts, theories and theme from this author. Read his introduction and conclusion and scan the rest. You now have the basis of your paper.

Look at the other books you have and use the other authors to support the ideas that you steal from "the book" and "the author." You have to have some quotes, so quote from these other tomes of wisdom and use the quotes that they use. (Only footnote them as if you had taken them from the original book—to show how extensively you have researched. Just be sure it is not a book impossible to obtain or originally written in early Swahili, the professor might cock an eye and say, "Hmmm.") It is not your fault that his particular thought was the most apt quote to expand what you were trying to say. (Obviously, great minds run along the same lines.) You can take up a lot of space by simply using very long quotes. Quoting from any author but "the author" is allowed.

Again, organize this paper along the lines we described earlier. The easiest way to do it is to divide the paper into ten-page sections. For example, if you need a forty-page paper, you will be writing four ten-page papers. You can discuss your main theory in the long introduction. Then go into three ten-page discussions on three different aspects of the theory. Your conclusion follows. Here is the part of the paper where you should be concise and very definite. If you need to stretch, go back to the introduction and add a paragraph or two on what drew you to this particular subject, or go back into the body of the paper and add a few more supporting quotes.

Bibliographies (which are often looked over scrupulously to see just how much work you did) are easily cribbed in such tight situations. Simply take two or three good books on the subject and use the books from their bibliographies. Just make sure that you do not use anything that is so old or so rare that the professor is going to wonder where you ever got hold of it. (Professors have been known to get so curious as to ask the student if he could get the copy for him to look at.) Your professor knows that you had fifteen weeks to do the thesis, but it is not likely that he will believe you made a special trip to the Library of Congress in Washington to look at this book, much less the British Museum or that you paid one and a half million for the only existing copy just to write your thesis. A good way to check yourself is to compare the bibliographies of your authors to see if the same books appear in several of them. If they are, you know that you are probably safe.

Do NOT use "the author" or "the book" in your bibliography.

Finally, your paper may not be long on originality, but you are bound to have a few ideas during the week that you are working on this. Work them in. Go to great lengths to explain them, but do not let them overwhelm the paper. Remember, you did—shall we say—"selective research." Your original theory could have been junked years ago, or worse yet, carefully thought out and explained fully by another author you have never heard of. So keep your enthusiasm under control.

Most important to remember: DO NOT PANIC. Keep your head together. You can do it. Just start writing. (Of course you have plenty of time if you do not intend to sleep, eat, go to the bathroom.)

HOW TO WRITE YOUR THESIS MORE QUICKLY: THE LOST WEEKEND

It is Friday morning and the birds are not singing but laughing at you because you have a thesis due on

Monday and you have not even begun. Take heart, it is not all over yet. (Get away from the window.) The worst thing you can do is sink into the depths of despair. Of course, now you regret leaving this momentous paper to the last minutes. But now that you have done it, there is no going back. (The platitude would be "There is no use crying over spilt milk," or "It is water under the bridge—(or over the dam.") The only thing to do is to forge ahead and write. Get cracking!

Since you have less than one hundred sixty-eight hours left, you are going to be able to do minimum research. Look for friends who took the course the year before. Call them up long distance, and if they are willing to part with it, take a day off to get their thesis, if necessary. No one throws away their senior thesis, so do not believe it if a "friend" tells you he does not have it anymore (Your thesis represents a kind of focal point for all you have done and suffered throughout your four years. To throw it away would be to say it was all for naught, it would be like tossing those four years out the window.) Twist a few arms.

Using someone else's thesis may sound like the audacity of the century to you. "Only someone with a death wish would try this," you shriek. But we say nay, you can use another person's thesis. (Granted, it will lack as much poignant meaning as an original, but it will suffice.) Unless you were a dead body languishing in a catatonic state in your classes during the past three years, you will have picked up enough knowledge to make this paper your own. Of course you will have to do a complete rewrite of the bloody thing, using a few of your own ideas, and adding a few of your own heavy-handed jokes properly stuffy and scholarly (English professors have a weakness for terrible puns). You should also write a new conclusion and introduction.

And presto!—instant thesis. The rewrite itself should take two days at the very most. The hard part will be the typing. Retyping the whole damn thing so that at least the typos will be original goes without saying.

If you do not have the guts or the necessary friend for the first solution, don't sweat it. You must have some of your old papers around—look at them. (We told you not to sell them—or at least, to keep copies.) There has to be one you could spend a day or two beefing up with quotes and expanding to the necessary depth and length. (This is one of the reasons we firmly believe that you should keep your footnotes as accurate as possible in all your papers. You never know.) Take this old paper and use it as the outline for your thesis. Now go deeply into each fact that you only mentioned in passing the last time around. Do not worry—this does not entail hours more of research. Get the books you used last time. Go through them and get quotes—long quotes. Quote everyone. Throw in a few more ideas of your own. The last professor to grade the paper made suggestions in the margins and possibly recommended books that you could have used in your bibliography. Do everything he suggested and quote the pithy books he recommended. (Think of it as a stew—you throw in everything you have at hand and hope.) Just keep writing.

Finally, if you do not have any guts, friends, or old papers to fall back on and you do have very little time, think of this story. One English/cinematography major in an Ivy League University in upstate New York, waited until the last minute to write her thesis in critical theory. This was really a panic situation, since many of the other students were preparing films to illustrate their critical theories. Our heroine did not know what to do. (This situation made *The Perils of Pauline* look like hayrides.) If she did not turn in a paper in five days, she would not graduate. Her parents had already made the hotel reservations.

This enterprising student did not give up. First she realized that the professor would probably die of boredom during the pretentious films made by the others. Also, fifty-page papers on the psychological effects of camera angles and lighting could not help but be boring after the first profundity (such as that a dark set creates overtones of anxiety touching off primal fears). She decided to be rather novel. (That was not an intentional pun.) She realized that she had to think of something that she knew well enough to write spontaneously on for fifty pages. She would have to have the books right at her fingertips. The only thing this student had right at her fingertips (besides chewed fingernails) were her books on Errol Flynn, the swashbuckling star of the 1930s. Having collected Flynn memorabilia for the last three years, she had a gold mine of information right in her room. She had also read about him during spare minutes in class, on the street, in coffee shops, etc., etc., and she had seen almost every movie he had made.

She wrote about Errol Flynn and the star system of the movie moguls in the 1930s. When this did not stretch out to the required page length, she padded it with a whimsical introduction about the great attraction for Errol Flynn she had always felt, and how the stars of the thirties exuded a great glamour, a mystique, etc. The mystery and magic movies, how modern critical theories could not be used to judge the older movies, and that old movies were really the prototype of audience manipulation was also part of her padding. She went on to say how she admired a man who seemed to smile his way through every adverse situation. (Rather apropos, actually.) She got an "A."

A bizarre case of luck, you might say. Untrue, we say. If you have spent three years majoring in something, then at some point, some aspect of it fascinated you. You probably looked into it on your own (because no one teaches courses on Flynn, Cagney, diseases, weirdos, etc.). But these subjects make the most interesting topics for papers. One public-health major at the University of Michigan hated the prospect of writing the thesis required of him. He got around it by writing a historical perspective on the effect of syphilis on history; focusing on the nineteenth century. When that did not reach the required word-count, he padded it with lengthy speculation about what might have happened if a few key figures of the period had not had the dread disease.

Another art major at Miami University wrote a paper on the anatomy dissections that Michaelangelo and his colleagues carried out all in the name of research. (Did you know that the best of society would come down and watch, that it was the social event of the season and that the skin was given to the reigning socialite of Florence?)

You see, you can do it. Just write about what you know best and work it around to fit into your class's theme or requirements. Show how it is background for the class, or a vital part, or a very interesting sidelight. Let your enthusiasm grow and blossom. Your enthusiasm—if it is really there—will carry the paper. Get yourself together. You know something about something. Even your obscure knowledge of subway systems or social relationships (you have a family, you have friends, you interrelate) or TV (if you are average, you watched enough television to write a book, much less a measly thesis). You can write a pretty damned interesting and good paper.

GREs, OTHER TESTS, AND GRADUATE
SCHOOL APPLICATIONS

You may have decided way back in your freshman year that you were definitely going to try and get into a graduate school. Now the time is here. You have to apply and take the proper tests and write out the applications. Just as in going to school and getting good grades, there are tricks to this trade.

As far as your applications go, there are deadlines as to when you must have them completed, including all records and recommendations. It is smart—extremely smart—to have them in as early as possible. The first year that the State University of New York at Stony Brook opened its medical school, they picked the lucky few by taking the most qualified people from the top of the pile (those applications that came in first). The rest of the people heard nothing for months, and then, just before the final term was over (and the due dates for most schools past), they got their rejections. Get your applications in early (and have backup schools).

As far as your recommendations go, build up a stockpile and keep them on file with the registrar. You do not want to go around asking for them a day before they are due. (Not only is one day's notice pressuring the professor, but chances are that he has been asked by a dozen people to write recommendations for them that night.) Ask every professor that you do get to know during your career to write one and hand it in to the registrar to add to your file. The advantage of doing this is fantastic. Under the new Freedom of Information law, the registrar must give you access to all and any files that they keep on you. If in the fall of your senior year you have a stockpile of recommendations in your file, you can ask the registrar to see them. Pick the best and ask that they be sent to your prospective schools. This is to your advantage. Previously you could have sent out a terrible recommendation, and never have been aware of it. Not to take advantage of this new law is sheer stupidity—and could be suicidal.

Most of the graduate-school applications have essay requirements. These can be a drag, but you must write them or your application will hit the round file. So when you are filling them out, be brief. Be very concise. Be heavy on scholarship and light on your good-hearted wit and humor. Individualistic oddballs were very "in" in the sixties, but they are definitely out now that there is a definite swing back to conservatism in the educational structure. So if you play in a banjo band, or study tombstones or fly kites with metal keys attached or skateboard on the freeway in your spare time, do not put it on your application. Your hobby is reading very obscure books in your field. Studying the early formation of your science or the development of ideas and theories in your field is your favorite pastime. But keeping up a very valuable collection of your field's older rare books is your real passion.

Many people feel the application essay should present the unblemished past of a Mr. Clean, no matter what the true situation is or what the accompanying transcripts reflect. Consequently, their essays make no mention of a catastrophic junior year which the transcript displays in burning, day-glo red. You have to use some judgment here. It goes without saying that you should not mention any smudges or disasters your transcripts and recommendations don't reveal. No one is going to find out that you are *the* speeding ticket outlaw of California or that you were busted with the rest of the fraternity for disorderly conduct—unless you tell them.

On the other hand, there are certain things you should explain. Transferring is no big deal, but you should throw in a sentence or two about why you did. (It has to be a pithy, valid sentence—not "I wanted to be with my boyfriend/girlfriend," even if that was the case.) If you made a radical change or transferred under somewhat cloudy circumstances, (i.e., you discovered you had arthritis and could no longer plan on a career as a professional musician, so you simply skipped all the tests and walked out of the conservatory one day—years later completing a B.A. in business), you ought to explain the transfer in greater depth than a sentence or two. If you don't, the admissions people are going to assume the worst. The worst can range from "This one can't make up his mind," to "Here's one who must have some really heavy emotional/ academic/legal/drug problems."

Aside from transferring, there is always a question of what to say about that one "disastrous" term or year when your grades just bottomed out. If it makes you feel any better, lots of people have this problem. Now if it was your freshman year, it's best not to mention it. Generally, the admissions people will chalk it up to "adjusting to college life." Anything beyond freshman year can be a problem. Obviously you can't say, "I was having such a dynamite time hanging out that I couldn't squeeze in any studying or classes." Also, if you are thinking of applying to professional schools, the last thing you want to mention is any type of emotional problem. (Even doctors get the blues, but no one talks about it.) The safest explanation is financial distress since poverty can strike even the sanest, soundest, and strongest of us all. Your term of C's and D's should be briefly covered as *the* grim moment of your life when you realized money was running out more quickly than you had anticipated. While attending school full time that year, you also worked thirty hours a week. Don't worry, no one is going to call the IRS and no grad school requires you to attach W-2s.

Finally, because so many people are applying to grad

schools, it is essential that you separate yourself from the teeming horde. The way to distinguish yourself on the application has changed throughout the years. Now you should write of your interest in one specific area of the chosen field, that one area that has gone out of style and been neglected lately. In other words, if you are applying to law school, don't just say you want to be a lawyer. There has been a drop in the number of students planning to work in Legal Aid programs. Legal Aid specialists are beginning to wilt in empty classrooms. No one will ever call you on it later and demand you turn in your shingle if you specialize in divorce law. (Unlike the Army, you do not have to sign a pledge promising the school a specific concentration.) It's just like applying to undergraduate school. Everyone puts down pre-law or pre-med. When the schools find these departments swamped, they start accepting more people applying for art history or Greek so that these departments won't feel neglected. Law, medical, and other graduate schools have to plan their class sizes like everyone else. A word of caution here—when you decide on a certain obscure, neglected interest, make sure that your prospective grad school has at least a course or two in it. You don't want to write reams about Egyptology for a history grad school that doesn't have at least one dusty professor writing on the subject. If you do, you will get a nice letter of regret suggesting a school that would be better suited to your interests. You are going to have to do a bit of research to discover each school's weak spot. Never fear, they exist; even institutions have Achilles heels.

THE TESTS

Just as the colleges demanded that you take the ACTs, SATs, and achievements, any graduate school is going to demand that you take the appropriate standardized exam. Let's not kid around here—these test scores are the determining factor of any graduate school application. Most schools claim they look at an application as a whole before rejecting it. This is a crock. They know it and you should know it. Despite all protests to the contrary, the scores are not used as indications of capability but as proof. As a result, the lawyers, doctors, and professionals of the future will definitely be the most talented test-takers; whether or not they will be even marginally talented lawyers, doctors, and professionals is another question.

Though grad schools deny it vociferously, they all have a cut-off point, and applications accompanied by test scores below that point are simply not looked at. (But they will open the envelope. No application fee check ever goes uncashed.) So if you score a 619 and send that score to a school with a 620 cut-off, rest assured that, unless you are a minority or equal opportunity student, your application will hit the floor and a rejection slip will follow shortly.

Knowing that your entire future hinges upon whether or not you fill in the computer dots correctly cannot help but make you a bit nervous. To make matters worse, if you do salvage the affair with a 700 the next time around, the first score is always there to remind the schools that you are either wildly erratic or had a lucky day. (They tend not to think of the first as a bad day.) Yes, it's still true: the testing services keep three scores on record and most schools have some bizarre, complex method of combining scores to get a "true" reading.

Needless to say, this is hardly fair. There are people with 3.99 averages who do not test well, and blank out when the tests come whistling across their desks. Aside from the nightmarish pressure that has been known to destroy a career or two, you have to wonder about the actual test contents: who or what are they testing? In one recent GMAT "event," reading selections included a passage by Jane Austen. Engineering students, scientists, and two English majors who had studied Austen with Lionel Trilling were evaluated on the basis of this test. Oh, ETS will say that this sort of thing evens out over the five sections of any test. Ha! Think of the poor engineer who gets four English sections and one Math; by the time this kid gets to the Math, he is experiencing academic meltdown. You might think things are changing. Under the "Truth in Testing" laws, Admissions Committees will have access to exams and will see just what they are using to evaluate students. Don't count on it. A simple cut-off point is very economically appealing to any Admissions office.

If you are enraged by this situation and feel that, for the fee you pay, your application should at least be looked at, fill out one of the minority student blanks. One senior we know of did just that. Several months later she was pleasantly surprised to receive a letter from a midwestern law school inviting her to submit an application, *and register*. They wanted to increase their enrollment of American Indians. This Irish Catholic from Long Island is now passing as an Algonquin somewhere in the Midwest.

Aside from blatantly lying about your origins, there are widely advertised "prep" schools and courses which claim they will help you raise your score at least a hundred points. Whether or not this is true, they are undeniably very expensive. Nevertheless, after January 1, 1982, copies of most grad exams and answer keys will be provided upon request to students tested. It is expected that these prep outfits will have their staffs tested several times a year. In the past, these people would try to memorize four or five questions for the practice exams they used to coach the "prepees." Now they will be able to request a copy of everything that isn't nailed down. Under copyright laws, they will not be allowed to give you a little booklet of previous tests. Let's just say that should you take one of these courses, you will probably have access to the most extensive collection of GMATs, LSATs, or GREs available. (If this doesn't bias the tests in favor of the more economically privileged, what does?)

This is not to say that such a course is essential. The people in Princeton who devise the boards claim that your current aptitudes will test out just about even with your SATs or ACTs no matter what you study or whom you study with. But if you just squeaked into college and are trembling at the thought of the grad boards you should look into one of these courses. You will learn the shortcuts and how to get the problems done in the time allotted. Most importantly, you will become familiar with the types of questions. Such familiarity is crucial. (People have been known to spend half the test time decoding the instructions.) After you take one of the prep courses, you may not know much more but you will be convinced that you do. Attitude alone tends to be worth about a hundred points. (Knowledge certainly isn't.)

If you can't afford a course, you should make every effort to start your own file of previous tests. (If you don't use it, you can always sell it.) If your friends will not request their graded exams, offer to do it for them. If they have requested them, beg for a copy. Obviously you are going to be at a disadvantage if you rely solely on practice exams in preparatory books. Of all these books, the *Barron's* provides the most comprehensive review of the material involved, but the practice tests *are not* actual ETS exams. It is expected that publishers will bring their practice exams more in line with reality as time goes on and more copies of ETS tests circulate. But right now they are not the real thing. And if, with a little effort you can get the real thing, you are a fool to pass it up.

Now if you are positive that this test is going to wipe out any career you might have, you can consider cheating. But remember, if they catch you, it's curtains,

Mugsy. You can't take the test ever again and no grad school is going to touch you. Besides, ordinary cheating (i.e., copying, coughing codes, passing answers) is simply impractical. At the test center, they have several different tests. The chances that you and your friend will have the same exam are minimal to nonexistent, as are the odds that you will be able to find someone to cheat with you. Crib notes might be useful if anyone had any idea of what to jot down. But the proctor puts a damper on this remaining possibility. ETS proctors come equipped with X-ray eyes. Logistically, do-it-yourself cheating is just impractical, unreliable, and extremely dangerous.

The only way to cheat on the grad boards is to pay someone to take the exam for you. By checking around, quietly, you can usually find someone with a proven track record of 600 or 700 who will do the dirty deed for a price, a stiff one. Before you invest, make sure this person is going to be able to deliver. (After all, if he fails to come through, you can hardly run off to Ralph Nader.) This "business associate" should be able to provide you with references. The going rate for warm bodies with brains is somewhere between seven hundred to one thousand dollars, but you can bargain. If you have a friend (a smart one, that is) who is willing to take them for you, dynamite.

With the increase in people paying others to take the exams, the testing services have taken precautions to ensure honesty (as opposed to justice). When entering the test, you will be asked to show identification including a physical description, such as a driver's license. In some cases, you will have to show a photo ID. Such ID can be obtained easily for your friend or business associate. Aside from mail order outfits advertising in teen fashion magazines, you can get one from your own friendly registrar. Have your employee go to the registrar and report that he (now you) has lost his ID and would like a replacement. After checking the student roster (which is by name not picture) they will send him over to the place where they take his photograph and put together a new ID. Give this person a few of your other vague IDs and you are all set.* Needless to say, it helps if this person looks a bit like you and is of the same sex.

Finally, on some of the exams you are required to fingerprint the answer sheet. This was supposed to eliminate the phonies for once and for all. It would have if everyone had to go into the Dean and match their fingerprints to the test in order for results to be recorded and forwarded. (But even then, what Dean knows every student by name? There would be nothing to prevent people from sending their employees— except the cost of keeping them in town until the results arrived.) This security precaution has not stopped the ringers. It's just made the process a little more time-consuming, difficult, and challenging.

The fingerprint will be used to verify your identity only if a question arises over the score you receive. If such a question does come up, you must report to an official of the test service accompanied by a person in a responsible position (usually your major advisor) who knows you and will swear to your identity. Then they take your fingerprint and match the two. This is where they catch you. (Of course, one day there may be a new breed of people: academic clones. They will assume your identity for your entire educational career, take all your tests, get all diplomas, and endure all the agony while the real you suns in Acapulco. No doubt such a luxury item would be prohibitively expensive, but what a great Christmas present!)

As we said earlier, it is possible to take the SAT sev-

---

* Any stranger claiming to be a professional test taker should have this procedure refined to an art. If you find yourself explaining how IDs will be obtained, you should look for another "pro." This one's 700 track record is probably as phony as his references.

eral times before you take it for the record book. Even under stricter grad board security procedures, this is still possible. You see, fingerprints are not cross-referenced by name. (Even the FBI has a hard time doing this.) So if you do not provide any information beyond a name, ETS can not keep continuing records on you. If by some remote chance they do manage to connect scores, simply call and inform them of the error: they have you mixed up with some other John Jones, Sally Spink, etc. If you have provided identifying information, you can take the test but request that it either not be graded or not sent to the schools listed. Once you request a test be scratched it cannot be reactivated and you lose all fees paid. But if you blank out completely or realize that the situation is totally hopeless, forfeit the fees. Why destroy a career for thirty bucks?

Remember, hiring someone is risky and very expensive. Before even looking for a "business associate," exploit the loopholes and see if things really are as bad as you believe. If the results indicate your budding career will be destroyed, you like the odds, and you have a good deal of sporting blood and spare cash, take a chance and look for a hired gun. If you're caught, well, chalk it up to the system, bad luck, or experience. If you had taken the test yourself, you wouldn't have made it either. So, what the hell, it's not a prison record.

GRAD SCHOOL: NOW, LATER, OR NEVER

A very strange thing happens during senior year: suddenly the end, the day you have been waiting for since they shoved you on the kindergarten bus, looms and you are scared. All of your classmates are freaking out. Everyone applies to graduate school. This is normal. Who knows what can happen out in the "real world"?

Aside from fear of the unknown, there is also a great deal of peer pressure and dazzle factor to deal with now. When everyone else is going somewhere, you feel as if you should too. If you haven't got a name to throw around, your peers may think you couldn't get in anywhere not that you didn't care to apply. And let's face it, being able to say "I'm going to El Supremo law school" is a lot more satisfying than "Zip Paper is starting me in the mailroom."

This is one of the most trying times of undergraduate school. It is very hard to ignore what everyone else is doing. But the graduate school decision is not a light one, and should not be made under duress, while intoxicated, or as an attempt to impress people. At the very least, further education is extremely expensive and time consuming. In addition, grad school is preparation for a specific career field: in most cases, what you prepare for is what you get.

Let's talk about this for a moment. Do you really know what you are preparing for? We are talking about the real grit of the job at the end of the rainbow. After sixteen years of schooling, you should know what teachers and professors do. Everyone knows what doctors do, so there is no problem there. But with business or law school, are you sure you know what you are getting into? Are you gathering your impressions from experience, or from newspapers and television?

Very few lawyers are actively involved in the hurly-burly of criminal work or politics. Many more specialize in litigation and corporate, family, or real estate law. Very few lawyers go on police chases or see the guilty break under relentless questioning in the courtroom. Before you zoom off to law school, take a paralegal or legal assistant's job. (You don't need any sort of training. A simple B.A. qualifies you.)

Obviously, such experience will expose you to the grit of the legal profession. You may find that it is not your cup of tea. Then again, you may discover that although real estate lawyers don't go on many stake-outs they do interesting work nevertheless. If this is the case, then make your interest known. Often your employers will write glowing recommendations and call old buddies who happen to be sitting on admissions committees. Once you are in law school you will be two steps ahead of the crowd in knowing what to specialize in and having connections in the field.

The case for taking time off, exploring, and gaining experience is even stronger in regard to business school. Ostensibly, B-schools provide the "managers of tomorrow" to industry. But very few industries require an MBA for entry level positions, and even fewer actually "need" MBAs on any level. Before you sink into debt for another twenty years, see if your area of interest has any interest in MBAs. If an industry does like MBAs, you still should not start sending for applications. Get a job with one of the bigger companies in the field instead. Why? Well, often if you show any potential at all, the company will not only recommend you to their favorite B-school they will finance your degree as well. We are talking about twenty grand here—isn't this worth working one year for peanuts?

Sure, we know that these entry level business slots aren't going to pay much (unlike the salaries reported for graduating MBAs). But—and this is a very big but—people with little or no experience do not get tremendous salaries anyway. Look at the average to lowest salaries listed in a B-school bulletin. That is where those without experience will start, unless they have undergraduate concentrations in such esoteric areas as petrochemical engineering. Average to upper-range salaries go to those with one or two years of experience. The people pulling down the astronomical starting salaries are the ones with experience in an industry, who went for the MBA, and are now returning to that field.

The same goes for law school. Unless you are at the top of the class, you are not going to pull down a truly stunning starting salary without some sort of experience (and summer jobs don't count). You can get all sorts of experience with summer internships but, in the long run, lawyers like to see some "real" experience. The average age at law and business schools is rising and those with little or no experience are at an increasing disadvantage when recruiting season rolls around.

You can see that it is definitely worthwhile to take time off between college and business or law school. The same holds true for just about any type of M.A. or Ph.D. program. Look at it this way: you have been in school for sixteen of your glorious twenty-one or twenty-two years. Why not check out the "real world" before you lock yourself away in academia? It might not be as bad as you think. Then again, it might. Either way you will find out, which is better than ending up as an unhappy professor wondering "If" at age thirty-five. Give it a shot. Expose yourself. At the very least, you will gain a broader perspective which will enable you to face the pressure of graduate or professional school with greater equanimity. Remember, the ivory tower will always welcome you back.

In certain circumstances, it is also wise to delay medical school. Obviously if you are accepted to any med school in the continental United States, you should immediately pack that microscope. But if you are accepted only overseas, you should carefully consider all the career implications. Many foreign schools are excellent and their students do practice in the U.S. Other schools exist simply to educate American rejectees and the long-term career prognosis is very doubtful. It might be more worthwhile to sit out a year, up your board scores, and re-apply to American schools.

This advice is all well and good, but you have already fallen prey to "grad school applicationitis," right? What if they accept you? How can you pass up such an opportunity? Well, you don't have to pass up anything. You simply request that your admission be deferred until the following fall. Yes, this can be done and it is done frequently. Deferred admissions are never rescinded; they will save a place for you one or two years down the road. You can't lose if you use this time to go out there and get some experience.

Graduate or professional school is not really going to prepare you for the specifics of any field. They aren't even going to teach you such specifics. Remember, school is school and graduate school just requires reading books with smaller print. But you can stack the deck in your favor by knowing exactly what you want when you go in. You will get a lot more out of it—and at these prices, you want to get everything that isn't nailed down.

FINANCIAL AID

There are so many places to get money to keep your academic career afloat that it is impossible to list them all or to even discuss the hundreds of *major* sources. The best we can do is offer some pointers on where to look for the cash, and how to keep your hot little hands on it after you get it.

Grants and Scholarships

Some experts have estimated that millions of dollars available for higher education go untapped each year. This situation may change as it gets harder to get federal education aid and loans, but at the moment, there is quite a bit of scholarship money sitting around gathering dust in trust funds throughout the country. Do you remember the rah-rah type we said would leave his estate to Alma Mater? Well, he did. But by the time he got around to making out the will, it seems our man was a little on the senile side. Consequently, there is a four-year college scholarship waiting for the student who is five foot two, wears size forty shoes, and comes from Alaska but lives in New Jersey now. The problem is that the Alaskan community in New Jersey is una-ware of this award, and so the small Alaskan with the big feet is not going to hear about it.

This problem has been solved to some degree by businesses such as Scholarship Search (1775 Broadway, New York, N.Y. 10019). This organization will send you an exhaustive questionnaire asking everything and anything that might possibly qualify you for aid. Then, for the minuscule price of $57 (or thereabouts), they will run your personal data through their super-duper computer which knows all about the hidden pockets of scholarship wealth nationwide. The Scholarship Search data bank lists over 250,000 different sources of financial aid—and this does not include state and federal sources! (Now you can see what the experts are talking about.) This particular outfit will find up to 25 sources you are qualified to apply for and subsequently exploit. If they don't find any, they will give your money back. But, they say, it is a rare student who comes up with zero on their computer.

Of course, in all cases, you still have to go through the application process and prove your worthiness. But, in the case of Scholarship Search, one out of every two students they handle receives aid from one or more of the sources they locate. Most of their applicants are advised of ten to fifteen sources worth $10,000 to $15,000. As you can see, it is well worth your time and effort to investigate. You can find out more about this organization and others like it by calling your local library or the college financial aid office. (Also, check with the Better Business Bureau before sending any money.) The testing services also run a similar search service which you can sign up for when you take the College Boards. But remember what we said about providing information to these organizations. Here again, your social security number is requested. So note that everything you reveal is filed in that big computer in the Jersey meadows. (Maybe if we all chipped in a buck, we could give a scholarship to the Alaskan New Jerseyite who can erect the most powerful magnet pointed towards Princeton. . . .)

Continuity is a major problem for many students on financial aid. Unfortunately, many of the tiny graduation prizes for poetry and the $250 grants (from the little old ladies of the Parsippany Society of the Lending Comforts to the Sick) tend to run out at the end of freshman year. This leaves you high and dry. You have to find other funds to keep your career going—or take the next train home. The financial aid office at your college can often be a great help if you suddenly find your well running dry. They will help locate new funds if you give them a chance and let them know that unless they do help, you are going to disappear. *But* you have to keep them advised of all changes in your financial position. They do not take kindly to people who walk in and announce they are broke the second week of the term. Or the people who wait until they do not have a full dollar to their names. It takes time to put together a financial boat; you risk drowning if you wait until the last minute.

If you are doing well (and you should be if you are following our advice), the financial aid office will tell you which of the university prizes and grants you are eligible to apply for. Many of these are given to promising students every year. Grants can range from a fifty dollar book award to next year's tuition. Apply for anything and everything you might possibly qualify for. It never hurts to try, and the worst they can say is "sorry."

Occasionally you can win a prize for poetry or journalism. Some schools give free dorm rooms to the editor of the campus newspaper, the student body president, or the captain of the football team (in the Midwest, it's the whole football team—which almost makes it worth joining). It is also common practice to give dorm counselors a free room, and sometimes a small monthly allowance. However, it's very obvious that a university needs only a limited number of captains, counselors and editors. The ordinary mortal must keep his eyes and ears alert to find out about university opportunities.

Remember, due to the cutbacks in federal education subsidies and loans, the competition for any and all types of grants and scholarships is very rough and is going to get much rougher. Stay awake!!

Loans

Aside from grants and scholarships, there are institutions which will loan you the money to expand your horizons. Among the most popular are the banks which lend money under the Federally Insured Student Loan Program. However, the government plans to make some changes in this program in the near future. At the very least, eligibility requirements for these loans are going to be tightened up. Also, it appears that the government subsidy of student loan interest is going to disappear—which means that students will borrow at current market interest rates, or slightly lower. Given tuition costs and the current interest rates, this prospect is truly frightening. If you are going to an expensive school these loans can run into the tens of thousands of dollars with more thousands in interest! You can see how easily a college education can put you in hock for the rest of your life. But face it, even if you are a total financial aid student, loans are inevitable. (No one escapes unscathed!)

Today most schools use a packaging system whereby they calculate your needs and then put together a "package" to meet them. This "package" includes all grants, scholarships and loans. So any money you come up with from various trust funds, grants and scholarships will go a long way toward easing the borrowed burden. (Then you will be out of hock at the age of fifty-nine—just in time to start saving for your retirement.)

Once you get grants or scholarships, many of them are good for the entire four years. Loans, however, usually have to be renewed annually (unless you took out one large loan and are merely using it in pieces). So it is essential, especially if your sources of income are scattered, that you know and meet all deadlines on the renewal applications. Almost every school requires the PCFS (Parents Confidential Financial Statement) be filed with the test services. (These people are everywhere!) Make sure you get it to your parents in time for them to fill it out and send it in. (And keep after them—parents have been known to forget.) Aside from doing your bit, you have to make sure the registrar is doing his. You should check, frequently, to make sure it doesn't take him three months to mail out your grades and "Student in Good Standing" form. (It may seem simple to look at a transcript and put an "x" in the appropriate box, but to the registrar this is an exacting task requiring months of exhausting effort.) Check. It's your neck, not his.

Finally, most schools require you maintain a specific point average to keep your aid (especially school grants). Often if your grade point falls below the required level, you are put on suspension until it is brought back up. Or worse yet, you may temporarily lose some or all of your aid. To a student on total financial aid, a temporary cut-off might as well be permanent. Either way, it spells the end.

Work/Study

Last—and sometimes least—are the work/study programs available on most campuses. The purpose is to provide students with eating money (though at today's prices, money earned will just about cover a diet of spaghetti). You have to prove financial need to get on work/study, which is a federal program whereby the government pays part, if not all, of your wages. You exhibit need with the PCFS in most cases.

If you know that you will need money for food and other necessities as soon as you get to campus, it is wise to sign up for work/study. You see, the program "guarantees" that the school will find you a job. For a freshman coming into a new town and making so many other adjustments, the worry of finding a job can be a tremendous burden. Let the school worry. They will find you a job in the dining halls, or doing a professor's clerical work, or shelving in the library. These jobs rarely pay more than the minimum wage and you are often allowed to work only so many hours a week. But if you count the hours of mental anguish you will go through before you find a job in a strange town during the first week of registration and classes, work/study is well worth it. Later, when you know your schedule and the way around town, you can look for a more lucrative position.

Finally, if you are an out-of-state student attending a state university, look into the state residency requirements. By fulfilling these requirements, you will become eligible for the instate tuition fees. In most cases, these are *substantially* lower than the fees for out-of-staters. The local Board of Elections or League of Women Voters can fill you in on the specifics of residency. Usually the financial aid office does not suggest or encourage this type of thing. It's not too hard to guess why.

You just have to keep your eyes and ears alert. There are ways, lots of them, to make money on campus, and many people who are willing to lend or give you money. All you have to do is find them. They don't hand it out on the street corners—you have to search them out. Get snooping, the sooner the better.

CONCLUSIONS: NOW YOU ARE READY TO FIGHT

When all is said and done, your transcripts, recommendations, and test scores will stand alone in front of a grad school admissions committee or potential employer. Nothing on your record is footnoted. No one will ever know that you deferred ten exams, never turned in a paper without an extension, or got your recommendations from a history prof who happens to be your best friend. We hope this manual has given you the street smarts to make that record look as good as it can under any light, no matter how harsh. What you have actually learned is something else entirely, and it doesn't matter to anyone but you. You see, no one expects you to "know" anything. They just want to see that little piece of paper covered with honors. Cynical, yes. But very, very true.

Now if you have a miserable time while landing every honor your university has to offer, you will miss one of the most important things college has to offer. At the risk of sounding like a guidance counselor, we must say that college is one of the finest moments of life. You can go to the movies for days, stay up all night, sleep all day, party and have a blast. Academia is an incubator sheltering you from the world of "real" responsibilities. You owe it to yourself to have the very best time possible.

Sure, those who follow our methods will not know as much as those who read every book mentioned in every class. Is this important? Not really. As we said, no grad school or employer expects you to know anything, in any depth. Besides, you have learned something that will be a hell of a lot more important in the long run: how to use a system to your advantage. "So what," you say. Good Lord! Listen, government is a system. Corporations are systems. Medicine is a system. Life is a system. Now are you properly in awe of what you have mastered?

When it's all over, you want to come out on top prepared to do anything and to go anywhere, with people waiting for *your* decision, not you waiting for theirs. If you know the system, you will walk down that graduation line with your classmates and you will be able to turn and wink:

MEUM HABEO: ET TU?*

* I got mine—how did you make out?